Recombinant DNA

Recombinant DNA

Amy Franko
Gary Thomas

Kruger Brentt
Publishers
2025

Kruger Brentt Publishers UK. LTD.
Company Number 9728962

Regd. Office: 68 St Margarets Road, Edgware, Middlesex HA8 9UU

ISBN: 978-1-78715-450-6

For information on all our publications visit our website at http://krugerbrentt.com/

PREFACE

Recombinant DNA (rDNA) technology is a groundbreaking scientific approach that involves the manipulation of genetic material to create new combinations of DNA. This process enables scientists to combine genes from different organisms, leading to the production of novel proteins or traits that would not naturally occur. Recombinant DNA technology has had a profound impact on various fields, including medicine, agriculture, and biotechnology, and has revolutionized the study of genetics.

The basic principle of recombinant DNA involves isolating a specific gene of interest from one organism and inserting it into a vector (such as a plasmid) to create a new DNA molecule. The vector is then introduced into a host organism, where it replicates and expresses the foreign gene. This allows researchers to produce large quantities of proteins, such as hormones or enzymes, or to study the function of a gene in a controlled environment. One of the most common tools in recombinant DNA technology is restriction enzymes, which act like molecular scissors to cut DNA at specific sequences. These enzymes enable scientists to isolate genes or modify DNA at precise locations. Another crucial tool is the ligase enzyme, which facilitates the joining of two pieces of DNA, allowing for the recombination of genetic material.

Recombinant DNA technology has numerous applications. In medicine, it has enabled the production of insulin for diabetes treatment, the creation of vaccines, and gene therapy for certain genetic disorders. In agriculture, it has led to the development of genetically modified (GM) crops with improved traits such as pest resistance or enhanced nutritional content. Furthermore, recombinant DNA technology plays a key role in scientific research by enabling the production of transgenic animals and plants, which serve as models for studying gene function and disease. Despite its immense benefits, recombinant DNA technology also raises ethical, safety, and environmental concerns, particularly regarding the use of genetically modified organisms (GMOs) in agriculture. The potential for unintended consequences, such as gene flow between GM and wild plants, is an area of ongoing research and debate.

Amy Franko | Gary Thomas

CONTENTS

PART 1:

FOUNDATIONS OF DNA

CHAPTER 1

DNA IS THE PRIMARY GENETIC MATERIAL

1.1 INTRODUCTION TO GENETIC MATERIAL

The search for the primary genetic material has been one of the most significant pursuits in the field of molecular biology. For centuries, scientists questioned how traits were passed from one generation to the next, speculating about the existence of a molecule that encodes genetic information. By the mid-20th century, researchers had identified deoxyribonucleic acid (DNA) as this molecule. DNA's structure, function, and properties make it the primary genetic material, responsible for storing, transmitting, and expressing genetic information essential for life.

1.2 HISTORICAL BACKGROUND AND DISCOVERY OF DNA'S ROLE

DNA was first isolated by the Swiss physician Friedrich Miescher who, in 1869, discovered a microscopic substance in the pus of discarded surgical bandages. As it resided in the nuclei of cells, he called it "nuclein". In 1878, Albrecht Kossel isolated the non-protein component of "nuclein", nucleic acid, and later isolated its five primary nucleobases.

In 1909, Phoebus Levene identified the base, sugar, and phosphate nucleotide unit of RNA (then named "yeast nucleic acid"). In 1929, Levene identified deoxyribose sugar in "thymus nucleic acid" (DNA). Levene suggested that DNA consisted of a string of four nucleotide units linked together through the phosphate groups ("tetranucleotide hypothesis"). Levene thought the chain was short and the bases repeated in a fixed order. In 1927, Nikolai Koltsov proposed that inherited traits would be inherited via a "giant hereditary molecule" made up of "two mirror strands that would replicate in a semi-conservative fashion using each strand as a template". In 1928, Frederick Griffith in his experiment discovered that traits of the "smooth" form of *Pneumococcus* could be transferred to the "rough" form of the same bacteria by mixing killed "smooth" bacteria with the live "rough" form. This system provided the first clear suggestion that DNA carries genetic information.

In 1933, while studying virgin sea urchin eggs, Jean Brachet suggested that DNA is found in the cell nucleus and that RNA is present exclusively in the cytoplasm.

At the time, "yeast nucleic acid" (RNA) was thought to occur only in plants, while "thymus nucleic acid" (DNA) only in animals. The latter was thought to be a tetramer, with the function of buffering cellular pH.

In 1937, William Astbury produced the first X-ray diffraction patterns that showed that DNA had a regular structure.

In 1943, Oswald Avery, along with co-workers Colin MacLeod and Maclyn McCarty, identified DNA as the transforming principle, supporting Griffith's suggestion (Avery–MacLeod–McCarty experiment). Erwin Chargaff developed and published observations now known as Chargaff's rules, stating that in DNA from any species of any organism, the amount of guanine should be equal to cytosine and the amount of adenine should be equal to thymine.

Late in 1951, Francis Crick started working with James Watson at the Cavendish Laboratory within the University of Cambridge. DNA's role in heredity was confirmed in 1952 when Alfred Hershey and Martha Chase in the Hershey–Chase experiment showed that DNA is the genetic material of the enterobacteria phage T2.

In May 1952, Raymond Gosling, a graduate student working under the supervision of Rosalind Franklin, took an X-ray diffraction image, labeled as "Photo 51", at high hydration levels of DNA. This photo was given to Watson and Crick by Maurice Wilkins and was critical to their obtaining the correct structure of DNA. Franklin told Crick and Watson that the backbones had to be on the outside. Before then, Linus Pauling, and Watson and Crick, had erroneous models with the chains inside and the bases pointing outwards.

Franklin's identification of the space group for DNA crystals revealed to Crick that the two DNA strands were antiparallel. In February 1953, Linus Pauling and Robert Corey proposed a model for nucleic acids containing three intertwined chains, with the phosphates near the axis, and the bases on the outside. Watson and Crick completed their model, which is now accepted as the first correct model of the double helix of DNA. On 28 February 1953 Crick interrupted patrons' lunchtime at The Eagle pub in Cambridge, England to announce that he and Watson had "discovered the secret of life".

The 25 April 1953 issue of the journal *Nature* published a series of five articles giving the Watson and Crick double-helix structure DNA and evidence supporting it. The structure was reported in a letter titled "*MOLECULAR STRUCTURE OF NUCLEIC ACIDS A Structure for Deoxyribose Nucleic Acid*", in which they said, "It has not escaped our notice that the specific pairing we have postulated immediately suggests a possible copying mechanism for the genetic material." This letter was followed by a letter from Franklin and Gosling, which was the first publication of their own X-ray diffraction data and of their original analysis method. Then followed a letter by Wilkins and two of his colleagues, which contained an analysis of *in vivo* B-DNA X-ray patterns, and which supported the presence *in vivo* of the Watson and Crick structure.

In April 2023, scientists, based on new evidence, concluded that Rosalind Franklin was a contributor and "equal player" in the discovery process of DNA, rather than otherwise, as may have been presented subsequently after the time of the discovery.

In 1962, after Franklin's death, Watson, Crick, and Wilkins jointly received the Nobel Prize in Physiology or Medicine. Nobel Prizes are awarded only to living recipients. A debate continues about who should receive credit for the discovery.

In an influential presentation in 1957, Crick laid out the central dogma of molecular biology, which foretold the relationship between DNA, RNA, and proteins, and articulated the "adaptor hypothesis". Final confirmation of the replication mechanism that was implied by the double-helical structure followed in 1958 through the Meselson–Stahl experiment. Further work by Crick and co-workers showed that the genetic code was based on non-overlapping triplets of bases, called codons, allowing Har Gobind Khorana, Robert W. Holley, and Marshall Warren Nirenberg to decipher the genetic code. These findings represent the birth of molecular biology.

In 1986, DNA analysis was first used for criminal investigative purposes when police in the UK requested Alec Jeffreys of the University of Leicester to verify or disprove a suspect's rape-murder "confession". In this particular case, the suspect had confessed to two rape-murders, but had later retracted his confession. DNA testing at the university labs soon disproved the veracity of the suspect's original "confession", and the suspect was exonerated from the murder-rape charges.

1.3 DNA STRUCTURE AND ITS IMPLICATIONS

The structure of DNA, elucidated by James Watson and Francis Crick in 1953, provided significant insights into how DNA functions as genetic material. DNA is composed of two strands arranged in a double helix, with each strand consisting of a sugar-phosphate backbone and nitrogenous bases (adenine, thymine, cytosine, and guanine) that pair specifically (adenine with thymine, and cytosine with guanine) through hydrogen bonding.

This structure allows DNA to replicate accurately, a critical feature for the transmission of genetic information from one cell generation to the next. The double helix model also provided a mechanism for genetic variability, as errors in replication or rearrangement of base pairs could result in mutations, leading to genetic diversity.

1.4 DNA REPLICATION AND GENETIC CONTINUITY

DNA's ability to replicate is a cornerstone of its role as the primary genetic material. During cell division, DNA replication ensures that each daughter cell receives an exact copy of the genetic material. The process is semiconservative, meaning that each new DNA molecule contains one original strand and one newly synthesized strand. Enzymes like DNA polymerase play essential roles in this process, ensuring that replication is both efficient and accurate. The fidelity of DNA replication is critical for maintaining genetic stability across generations, and any errors that escape correction can lead to mutations. This process highlights DNA's unique suitability as the genetic material, as it ensures continuity and stability of genetic information.

1.5 DNA AND GENE EXPRESSION

One of the most critical aspects of DNA as genetic material is its role in gene expression. DNA does not directly participate in cellular activities; instead, it acts as a template to create messenger RNA (mRNA), which carries genetic instructions to the ribosomes for protein synthesis. This process, known as transcription, involves copying a gene's sequence into mRNA. The mRNA then undergoes translation, where ribosomes synthesize proteins based on the sequence of nucleotides in the mRNA. Proteins are vital to cellular function, as they perform structural, catalytic, and regulatory roles. Thus, DNA indirectly determines the traits and functions of an organism by encoding the structure of proteins, further establishing its role as the primary genetic material.

1.6 DNA'S STABILITY AND ADAPTABILITY

The chemical structure of DNA lends it a high level of stability, which is essential for long-term genetic storage. The double-stranded nature of DNA protects it from enzymatic degradation, while the helical structure shields it from potential damage. Additionally, the redundancy of base pairing provides a mechanism for repair, as damaged or mismatched bases can often be corrected by cellular repair mechanisms. Despite its stability, DNA is not entirely immutable; it can undergo mutations, either through errors in replication or in response to environmental factors such as radiation and chemicals. These mutations can result in genetic variation, which drives evolution and adaptation. DNA's combination of stability and adaptability makes it well-suited to serve as the primary genetic material over the long term.

1.7 DNA IN HEREDITY AND EVOLUTION

DNA's role in heredity is central to the survival and evolution of species. Through processes like recombination during meiosis, DNA allows for the reshuffling of genes, leading to unique combinations of traits in offspring. This recombination contributes to genetic diversity within a population, increasing its adaptability to changing environments. Additionally, mutations in DNA can introduce new traits, some of which may offer survival advantages. These advantageous mutations may become more common in a population over time through natural selection. Therefore, DNA not only enables inheritance but also plays a crucial role in evolution, providing the raw material for natural selection to act upon.

1.8 DNA IN MODERN GENETICS AND BIOTECHNOLOGY

The recognition of DNA as the primary genetic material has revolutionized the fields of genetics and biotechnology. Techniques such as polymerase chain reaction (PCR), gene sequencing, and gene editing rely on the understanding of DNA's structure and function. PCR allows scientists to amplify specific DNA sequences, making it possible to study genes in detail, diagnose diseases, and conduct forensic investigations. Gene sequencing has enabled researchers to decode entire genomes, offering insights into genetic disorders, evolution, and population genetics. Furthermore, gene-editing technologies like CRISPR-Cas9 enable precise modifications to DNA, with applications in medicine, agriculture, and environmental science. These advancements underscore DNA's central role as the primary genetic material and its vast potential in scientific research and applications.

1.9 DNA AND THE CENTRAL DOGMA OF MOLECULAR BIOLOGY

The central dogma of molecular biology, proposed by Francis Crick, describes the flow of genetic information from DNA to RNA to protein. This model underscores the role of DNA as the repository of genetic information, with RNA serving as a messenger and protein as the functional output. The central dogma illustrates how genetic information encoded in DNA is ultimately expressed as cellular traits and functions. It also emphasizes the unidirectional flow of information, highlighting DNA's role as the origin and storage point of genetic material in cells. Understanding this process has been essential for research into genetic regulation, mutations, and hereditary diseases.

1.10 CONCLUSION

DNA's discovery and its role as the primary genetic material have transformed our understanding of life. Its unique properties—replication, gene expression, stability, and adaptability—make it the ideal molecule for storing and transmitting genetic information across generations. From basic cellular processes to complex evolutionary adaptations, DNA is integral to all life forms. Modern biotechnological advancements continue to leverage DNA's properties, opening new avenues in medicine, agriculture, and environmental conservation. As we further explore DNA's potential, it remains at the heart of genetic research and applications, continually reaffirming its role as the blueprint of life.

CHAPTER 2

INFORMATION FLOW FROM DNA TO PROTEIN

2.1 INTRODUCTION TO THE CENTRAL DOGMA OF MOLECULAR BIOLOGY

The concept of information flow from DNA to protein is central to molecular biology. This process, often referred to as the "central dogma" of molecular biology, describes the sequence of events that lead to the synthesis of proteins from genetic information encoded in DNA. Proposed by Francis Crick in 1958, the central dogma outlines the transfer of information in a unidirectional path: DNA to RNA to protein. This pathway involves two primary stages—transcription and translation—that together govern how genetic instructions are transformed into functional proteins. Proteins, in turn, play crucial roles in cellular structure, function, and regulation, making the flow of information from DNA to protein essential for life.

2.2 STRUCTURE OF DNA AND GENETIC CODING

DNA (deoxyribonucleic acid) is a double-helical molecule composed of nucleotide building blocks, each containing a sugar, a phosphate group, and one of four nitrogenous bases: adenine, thymine, cytosine, and guanine. The sequence of these bases encodes the genetic instructions for building proteins. Genes are specific sequences of DNA that correspond to particular proteins. Each gene comprises a series of triplets, or codons, each coding for a specific amino acid. The order of these codons determines the sequence of amino acids in a protein, thereby dictating its structure and function. Understanding the organization of genetic information within DNA is the first step in understanding the process by which this information is converted into proteins.

2.3 TRANSCRIPTION: THE FIRST STEP IN GENE EXPRESSION

Transcription is the process of copying a segment of RNA into DNA. Some segments of DNA are transcribed into RNA molecules that can encode proteins, called messenger RNA (mRNA). Other segments of DNA are transcribed into RNA molecules called non-coding RNAs (ncRNAs). Both DNA and RNA are nucleic acids, which use base pairs of nucleotides as a complementary language. During transcription, a DNA sequence is read by an RNA polymerase, which produces a complementary, antiparallel RNA strand called a primary transcript. In virology, the term transcription is used when referring to mRNA synthesis from a viral RNA molecule. The genome of many RNA viruses is composed of negative-sense RNA which acts as a template for positive sense viral messenger RNA - a necessary step in the synthesis of viral proteins needed for viral replication. This process is catalyzed by a viral RNA dependent RNA polymerase.

2.3.1 Background

A DNA transcription unit encoding for a protein may contain both a *coding sequence*, which will be translated into the protein, and *regulatory sequences*, which direct and regulate the synthesis of that protein. The regulatory sequence before (upstream from) the coding sequence is called the five prime untranslated regions (5'UTR); the sequence after (downstream from) the coding sequence is called the three prime untranslated regions (3'UTR).

As opposed to DNA replication, transcription results in an RNA complement that includes the nucleotide uracil (U) in all instances where thymine (T) would have occurred in a DNA complement.

Only one of the two DNA strands serves as a template for transcription. The antisense strand of DNA is read by RNA polymerase from the 3' end to the 5' end during transcription (3' → 5'). The complementary RNA is created in the opposite direction, in the 5' → 3' direction, matching the sequence of the sense strand except switching uracil for thymine. This directionality is because RNA polymerase can only add nucleotides to the 3' end of the growing mRNA chain. This use of only the 3' → 5' DNA strand eliminates the need for the Okazaki fragments that are seen in DNA replication. This also removes the need for an RNA primer to initiate RNA synthesis, as is the case in DNA replication.

The *non*-template (sense) strand of DNA is called the coding strand, because its sequence is the same as the newly created RNA transcript (except for the substitution of uracil for thymine). This is the strand that is used by convention when presenting a DNA sequence. Transcription has some proofreading mechanisms, but they are fewer and less effective than the controls for copying DNA. As a result, transcription has a lower copying fidelity than DNA replication.

2.3.2 Major Steps

Setting up for Transcription

Enhancers, Transcription Factors, Mediator Complex, and DNA Loops in Mammalian Transcription

Setting up for transcription in mammals is regulated by many cis-regulatory elements, including core promoter and promoter-proximal elements that are located near the transcription start sites of genes. Core promoters combined with general transcription factors are sufficient to direct transcription initiation, but generally have low basal activity. Other important cis-regulatory modules are localized in DNA regions that are distant from the transcription start sites. These include enhancers, silencers, insulators and tethering elements. Among this constellation of elements, enhancers and their associated transcription factors have a leading role in the initiation of gene transcription. An enhancer localized in a DNA region distant from the promoter of a gene can have a very large effect on gene transcription, with some genes undergoing up to 100-fold increased transcription due to an activated enhancer.

Enhancers are regions of the genome that are major gene-regulatory elements. Enhancers control cell-type-specific gene transcription programs, most often by looping through long distances to come in physical proximity with the promoters of their target genes. While there are hundreds of thousands of enhancer DNA regions, for a particular type of tissue only specific enhancers are brought into proximity with the promoters that they regulate. In a study of brain cortical neurons, 24,937 loops were found, bringing enhancers to their target promoters. Multiple enhancers, each often at tens or hundred of thousands of nucleotides distant from their target genes, loop to their target gene promoters and can coordinate with each other to control transcription of their common target gene.

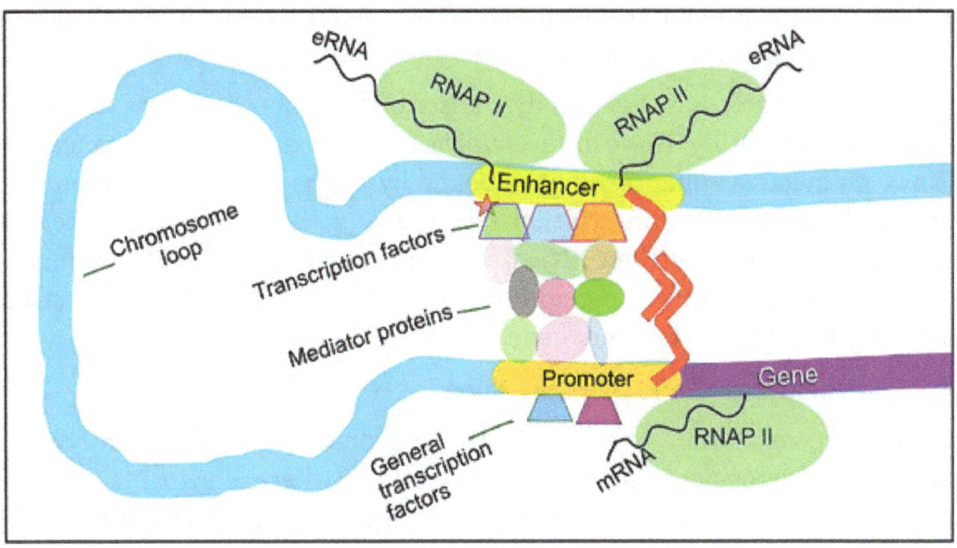

Fig: Regulation of Transcription in Mammals

An active enhancer regulatory region of DNA is enabled to interact with the promoter DNA region of its target gene by the formation of a chromosome loop. This can initiate messenger RNA (mRNA) synthesis by RNA polymerase II (RNAP

II) bound to the promoter at the transcription start site of the gene. The loop is stabilized by one architectural protein anchored to the enhancer and one anchored to the promoter and these proteins are joined to form a dimer (red zigzags). Specific regulatory transcription factors bind to DNA sequence motifs on the enhancer. General transcription factors bind to the promoter. When a transcription factor is activated by a signal (here indicated as phosphorylation shown by a small red star on a transcription factor on the enhancer) the enhancer is activated and can now activate its target promoter. The active enhancer is transcribed on each strand of DNA in opposite directions by bound RNAP IIs. Mediator (a complex consisting of about 26 proteins in an interacting structure) communicates regulatory signals from the enhancer DNA-bound transcription factors to the promoter.

The loop is stabilized by a dimer of a connector protein (e.g., dimer of CTCF or YY1), with one member of the dimer anchored to its binding motif on the enhancer and the other member anchored to its binding motif on the promoter (represented by the red zigzags in the illustration). Several cell function specific transcription factors (there are about 1,600 transcription factors in a human cell) generally bind to specific motifs on an enhancer and a small combination of these enhancer-bound transcription factors, when brought close to a promoter by a DNA loop, govern level of transcription of the target gene. Mediator (a complex usually consisting of about 26 proteins in an interacting structure) communicates regulatory signals from enhancer DNA-bound transcription factors directly to the RNA polymerase II (pol II) enzyme bound to the promoter.

Enhancers, when active, are generally transcribed from both strands of DNA with RNA polymerases acting in two different directions, producing two enhancer RNAs (eRNAs). An inactive enhancer may be bound by an inactive transcription factor. Phosphorylation of the transcription factor may activate it and that activated transcription factor may then activate the enhancer to which it is bound. An activated enhancer begins transcription of its RNA before activating transcription of messenger RNA from its target gene.

CpG Island Methylation and Demethylation

Transcription regulation at about 60% of promoters is also controlled by methylation of cytosines within CpG dinucleotides (where 5' cytosine is followed by 3' guanine or CpG sites). 5-methylcytosine (5-mC) is a methylated form of the DNA base cytosine. 5-mC is an epigenetic marker found predominantly within CpG sites. About 28 million CpG dinucleotides occur in the human genome. In most tissues of mammals, on average, 70% to 80% of CpG cytosines are methylated (forming 5-methylCpG or 5-mCpG). However, unmethylated cytosines within 5'cytosine-guanine 3' sequences often occur in groups, called CpG islands, at active promoters. About 60% of promoter sequences have a CpG island while only about

6% of enhancer sequences have a CpG island. CpG islands constitute regulatory sequences, since if CpG islands are methylated in the promoter of a gene this can reduce or silence gene transcription.

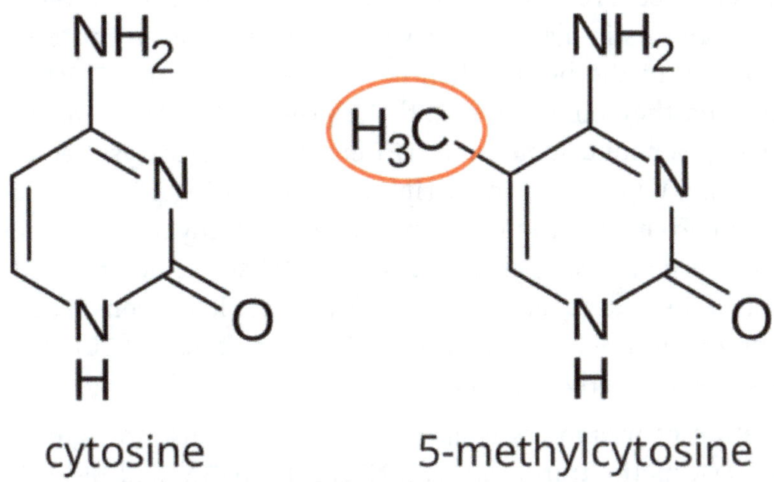

Fig: This shows where the Methyl Group is Added when 5-Methylcytosine is Formed

DNA methylation regulates gene transcription through interaction with methyl binding domain (MBD) proteins, such as MeCP2, MBD1 and MBD2. These MBD proteins bind most strongly to highly methylated CpG islands. These MBD proteins have both a methyl-CpG-binding domain as well as a transcription repression domain. They bind to methylated DNA and guide or direct protein complexes with chromatin remodeling and/or histone modifying activity to methylated CpG islands. MBD proteins generally repress local chromatin such as by catalyzing the introduction of repressive histone marks, or creating an overall repressive chromatin environment through nucleosome remodeling and chromatin reorganization.

Transcription factors are proteins that bind to specific DNA sequences in order to regulate the expression of a gene. The binding sequence for a transcription factor in DNA is usually about 10 or 11 nucleotides long. As summarized in 2009, Vaquerizas et al. indicated there are approximately 1,400 different transcription factors encoded in the human genome by genes that constitute about 6% of all human protein encoding genes. About 94% of transcription factor binding sites (TFBSs) that are associated with signal-responsive genes occur in enhancers while only about 6% of such TFBSs occur in promoters.

EGR1 protein is a particular transcription factor that is important for regulation of methylation of CpG islands. An EGR1 transcription factor binding site is frequently located in enhancer or promoter sequences. There are about 12,000 binding sites for EGR1 in the mammalian genome and about half of EGR1 binding sites are located in promoters and half in enhancers. The binding of EGR1 to its target DNA binding site is insensitive to cytosine methylation in the DNA. While only small amounts of EGR1 transcription factor protein are detectable in cells that are unstimulated, translation of the *EGR1* gene into protein at one hour after stimulation is drastically elevated. Production of EGR1 transcription factor proteins, in various types of cells, can be stimulated by growth factors, neurotransmitters, hormones, stress and injury. In the brain, when neurons are activated, EGR1 proteins are up-regulated and they bind to (recruit) the pre-existing TET1 enzymes that are produced in high amounts in neurons. TET enzymes can catalyse demethylation of 5-methylcytosine. When EGR1 transcription factors bring TET1 enzymes to EGR1 binding sites in promoters, the TET enzymes can demethylate the methylated CpG islands at those promoters. Upon demethylation, these promoters can then initiate transcription of their target genes. Hundreds of genes in neurons are differentially expressed after neuron activation through EGR1 recruitment of TET1 to methylated regulatory sequences in their promoters.

The methylation of promoters is also altered in response to signals. The three mammalian DNA methyltransferasess (DNMT1, DNMT3A, and DNMT3B) catalyze the addition of methyl groups to cytosines in DNA. While DNMT1 is a maintenance methyltransferase, DNMT3A and DNMT3B can carry out new methylations. There are also two splice protein isoforms produced from the *DNMT3A* gene: DNA methyltransferase proteins DNMT3A1 and DNMT3A2.

The splice isoform DNMT3A2 behaves like the product of a classical immediate-early gene and, for instance, it is robustly and transiently produced after neuronal activation. Where the DNA methyltransferase isoform DNMT3A2 binds and adds methyl groups to cytosines appears to be determined by histone post translational modifications.

On the other hand, neural activation causes degradation of DNMT3A1 accompanied by reduced methylation of at least one evaluated targeted promoter.

Initiation

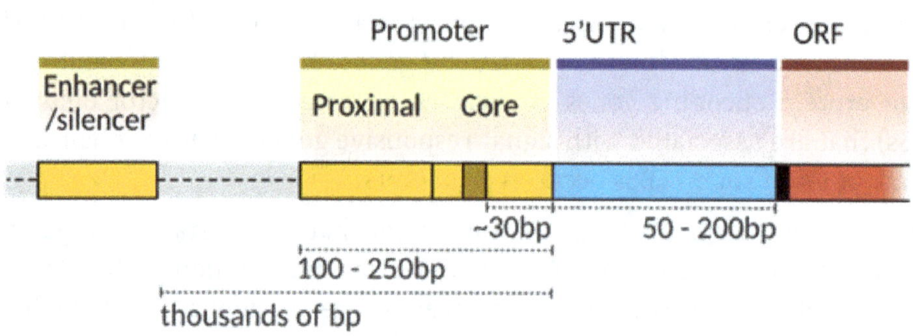

Fig: The Regulatory Sequence Elements

The regulatory sequence elements (yellow) at the start of a eukaryotic protein-coding gene, can be immediately upstream of the open read frame (ORF, red), or many kilobases away (upstream or downstream). Promoter and enhancer regions up-regulate (and silencers downregulate) transcription from DNA to mRNA. The 5' and 3' untranslated regions of that mRNA (UTR, blue) then regulate translation into the final protein product.

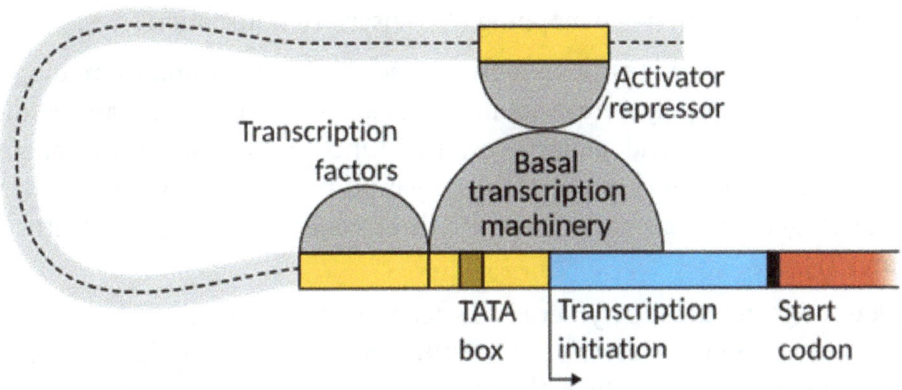

Fig: During Transcription Initiation

During transcription initiation, proteins (dark grey semi-circles) bound to the DNA can be brought into proximity with each other since the intervening DNA can loop back on itself. In this way, the basal transcription machinery can interact with distant activators and repressors many kilobases upstream or downstream of the open reading frame.

Transcription begins with the RNA polymerase and one or more general transcription factors binding to a DNA promoter sequence to form an RNA polymerase-promoter closed complex. In the closed complex, the promoter DNA is still fully double-stranded.

RNA polymerase, assisted by one or more general transcription factors, then unwinds approximately 14 base pairs of DNA to form an RNA polymerase-promoter open complex. In the open complex, the promoter DNA is partly unwound and single-stranded. The exposed, single-stranded DNA is referred to as the "transcription bubble". RNA polymerase, assisted by one or more general transcription factors, then selects a transcription start site in the transcription bubble, binds to an initiating NTP and an extending NTP (or a short RNA primer and an extending NTP) complementary to the transcription start site sequence, and catalyzes bond formation to yield an initial RNA product.

In bacteria, RNA polymerase holoenzyme consists of five subunits: 2 α subunits, 1 β subunit, 1 β' subunit, and 1 ω subunit. In bacteria, there is one general RNA transcription factor known as a sigma factor. RNA polymerase core enzyme binds to the bacterial general transcription (sigma) factor to form RNA polymerase holoenzyme and then binds to a promoter. (RNA polymerase is called a holoenzyme when sigma subunit is attached to the core enzyme which is consist of 2 α subunits, 1 β subunit, 1 β' subunit only). Unlike eukaryotes, the initiating nucleotide of nascent bacterial mRNA is not capped with a modified guanine nucleotide. The initiating nucleotide of bacterial transcripts bears a 5' triphosphate (5'-PPP), which can be used for genome-wide mapping of transcription initiation sites.

In archaea and eukaryotes, RNA polymerase contains subunits homologous to each of the five RNA polymerase subunits in bacteria and also contains additional subunits. In archaea and eukaryotes, the functions of the bacterial general transcription factor sigma are performed by multiple general transcription factors that work together. In archaea, there are three general transcription factors: TBP, TFB, and TFE.

In eukaryotes, in RNA polymerase II-dependent transcription, there are six general transcription factors: TFIIA, TFIIB (an ortholog of archaeal TFB), TFIID (a multisubunit factor in which the key subunit, TBP, is an ortholog of archaeal TBP), TFIIE (an ortholog of archaeal TFE), TFIIF, and TFIIH. The TFIID is the first component to bind to DNA due to binding of TBP, while TFIIH is the last component to be recruited. In archaea and eukaryotes, the RNA polymerase-promoter closed complex is usually referred to as the "preinitiation complex".

Transcription initiation is regulated by additional proteins, known as activators and repressors, and, in some cases, associated coactivators or corepressors, which modulate formation and function of the transcription initiation complex.

Promoter Escape

After the first bond is synthesized, the RNA polymerase must escape the promoter. During this time there is a tendency to release the RNA transcript and produce truncated transcripts. This is called abortive initiation, and is common for both eukaryotes and prokaryotes. Abortive initiation continues to occur until an RNA product of a threshold length of approximately 10 nucleotides is synthesized, at which point promoter escape occurs and a transcription elongation complex is formed.

Mechanistically, promoter escape occurs through DNA scrunching, providing the energy needed to break interactions between RNA polymerase holoenzyme and the promoter.

In bacteria, it was historically thought that the sigma factor is definitely released after promoter clearance occurs. This theory had been known as the *obligate release model.* However, later data showed that upon and following promoter clearance, the sigma factor is released according to a stochastic model known as the *stochastic release model.*

In eukaryotes, at an RNA polymerase II-dependent promoter, upon promoter clearance, TFIIH phosphorylates serine 5 on the carboxy terminal domain of RNA polymerase II, leading to the recruitment of capping enzyme (CE). The exact mechanism of how CE induces promoter clearance in eukaryotes is not yet known.

Elongation

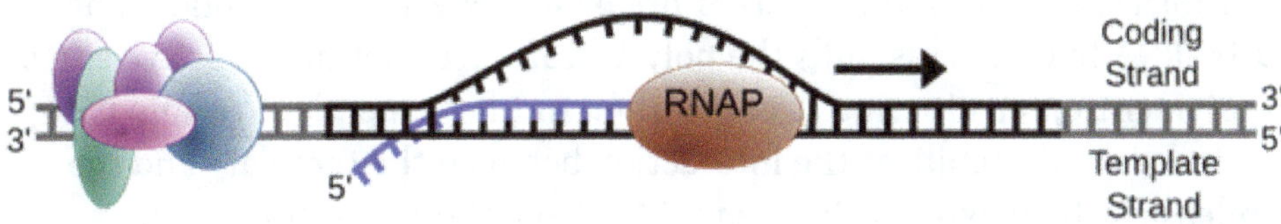

Fig: Simple Diagram of Transcription Elongation

One strand of the DNA, the *template strand* (or noncoding strand), is used as a template for RNA synthesis. As transcription proceeds, RNA polymerase traverses the template strand and uses base pairing complementarity with the DNA template to create an RNA copy (which elongates during the traversal). Although RNA polymerase traverses the template strand from 3' → 5', the coding (non-template) strand and newly formed RNA can also be used as reference points, so transcription can be described as occurring 5' → 3'. This produces an RNA molecule from 5' → 3', an exact copy of the coding strand (except that thymines are replaced with uracils, and the nucleotides are composed of a ribose (5-carbon) sugar whereas DNA has deoxyribose (one fewer oxygen atom) in its sugar-phosphate backbone).

mRNA transcription can involve multiple RNA polymerases on a single DNA template and multiple rounds of transcription (amplification of particular mRNA), so many mRNA molecules can be rapidly produced from a single copy of a gene. The characteristic elongation rates in prokaryotes and eukaryotes are about 10–100 nts/sec. In eukaryotes, however, nucleosomes act as major barriers to transcribing polymerases during transcription elongation. In these organisms, the pausing induced by nucleosomes can be regulated by transcription elongation factors such as TFIIS.

Elongation also involves a proofreading mechanism that can replace incorrectly incorporated bases. In eukaryotes, this may correspond with short pauses during transcription that allow appropriate RNA editing factors to bind. These pauses may be intrinsic to the RNA polymerase or due to chromatin structure.

Double-strand breaks in actively transcribed regions of DNA are repaired by homologous recombination during the S and G2 phases of the cell cycle. Since transcription enhances the accessibility of DNA to exogenous chemicals and internal metabolites that can cause recombinogenic lesions, homologous recombination of a particular DNA sequence may be strongly stimulated by transcription.

Termination

Bacteria use two different strategies for transcription termination - Rho-independent termination and Rho-dependent termination. In Rho-independent transcription termination, RNA transcription stops when the newly synthesized RNA molecule forms a G-C-rich hairpin loop followed by a run of Us. When the hairpin forms, the mechanical stress breaks the weak rU-dA bonds, now filling the DNA–RNA hybrid. This pulls the poly-U transcript out of the active site of the RNA polymerase, terminating transcription. In Rho-dependent termination, Rho, a protein factor, destabilizes the interaction between the template and the mRNA, thus releasing the newly synthesized mRNA from the elongation complex.

Transcription termination in eukaryotes is less well understood than in bacteria, but involves cleavage of the new transcript followed by template-independent addition of adenines at its new 3' end, in a process called polyadenylation.

Beyond termination by a terminator sequences (which is a part of a gene), transcription may also need to be terminated when it encounters conditions such as DNA damage or an active replication fork. In bacteria, the Mfd ATPase can remove a RNA polymerase stalled at a lesion by prying open its clamp. It also recruits nucleotide excision repair machinery to repair the lesion. Mfd is proposed to also resolve conflicts between DNA replication and transcription. In eukayrotes, ATPase TTF2 helps to suppress the action of RNAP I and II during mitosis, preventing errors in chromosomal segregation. In archaea, the Eta ATPase is proposed to play a similar role.

2.3.3 Inhibitors

Transcription inhibitors can be used as antibiotics against, for example, pathogenic bacteria (antibacterials) and fungi (antifungals). An example of such an antibacterial is rifampicin, which inhibits bacterial transcription of DNA into mRNA by inhibiting DNA-dependent RNA polymerase by binding its beta-subunit, while 8-hydroxyquinoline is an antifungal transcription inhibitor. The effects of histone methylation may also work to inhibit the action of transcription. Potent, bioactive natural products like triptolide that inhibit mammalian transcription via inhibition of the XPB subunit of the general transcription factor TFIIH has been recently reported as a glucose conjugate for targeting hypoxic cancer cells with increased glucose transporter production.

2.3.4 Endogenous Inhibitors

In vertebrates, the majority of gene promoters contain a CpG island with numerous CpG sites. When many of a gene's promoter CpG sites are methylated the gene becomes inhibited (silenced). Colorectal cancers typically have 3 to 6 driver mutations and 33 to 66 hitchhiker or passenger mutations. However, transcriptional inhibition (silencing) may be of more importance than mutation in causing progression to cancer. For example, in colorectal cancers about 600 to 800 genes are transcriptionally inhibited by CpG island methylation. Transcriptional repression in cancer can also occur by other epigenetic mechanisms, such as altered production of microRNAs. In breast cancer, transcriptional repression of BRCA1 may occur more frequently by over-produced microRNA-182 than by hypermethylation of the BRCA1 promoter.

2.3.5 Transcription Factories

Transcription factories, in genetics describe the discrete sites where transcription occurs in the cell nucleus, and are an example of a biomolecular condensate.

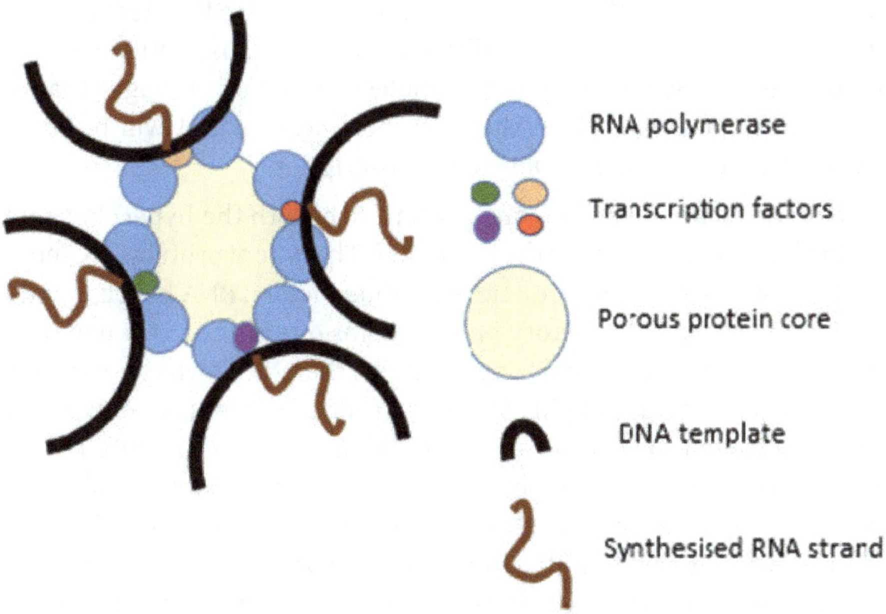

Fig: A Generic Transcription Factory during Transcription, Highlighting the Possibility of Transcribing more than one Gene at a Time. The Diagram includes 8 RNA Polymerases, however, the Number can vary Depending on Cell Type. The Image also includes Transcription Factors and a Porous, Protein Core

They were first discovered in 1993 and have been found to have structures analogous to replication factories, sites where replication also occurs in discrete sites. The factories contain an RNA polymerase (active or inactive) and the necessary transcription factors (activators and repressors) for transcription. Transcription factories containing RNA polymerase II are the most studied but factories can exist for RNA polymerase I and III; the nucleolus being seen as the prototype for transcription factories. It is possible to view them under both light and electron microscopy. The discovery of transcription factories has challenged the original view of how RNA polymerase interacts with the DNA polymer and it is thought that the presence of factories has important effects on gene regulation and nuclear structure.

Discovery

The first use of the term 'transcription factory' was used in 1993 by *Jackson* and his colleagues who noticed that transcription occurred at discrete sites in the nucleus. This contradicted the original view that transcription occurred at an even distribution throughout the nucleus.

Structure

The structure of a transcription factory appears to be determined by cell type, transcriptional activity of the cell and also the method of technique used to visualise the structure. The generalised view of a transcription factory would feature between 4 – 30 RNA polymerase molecules and it is thought that the more transcriptionally active a cell is, the more polymerases that will be present in a factory in order to meet the demands of transcription.

The core of the factory is porous and protein rich, with the hyperphosphorylated, elongating form polymerases on the perimeter. The type of proteins present include: ribonucleoproteins, co-activators, transcription factors, RNA helicase and splicing and processing enzymes. A factory only contains one type of RNA polymerase and the diameter of the factory varies depending on the RNA polymerase featured; RNA polymerase I factories are roughly 500 nm in width whereas RNA polymerase II and III factories a magnitude smaller at 50 nm. It has been experimentally shown that the transcription factory is immobilised to a structure and it is postulated that this immobilisation is because of a tethering to the nuclear matrix; this is because it has been shown it is tied to a structure that is unaffected by restriction enzymes. Proteins that have been thought to be involved in the tethering includes spectrin, actin and lamins.

Function

The structure of a transcriptional factory directly relates to its function. Transcription is made more efficient because of the clustered nature of the transcription factory. All the necessary proteins: RNA polymerase, transcription factors and other co-regulators are present in the transcription factory that allows for faster RNA polymerisation when the DNA template reaches the factory, it also allows for a number of genes to be transcribed at the same time.

Genomic Location

The amount of transcription factories found per nucleus appears to be determined by cell type, species and the type of measurement. Cultured mouse embryonic fibroblasts have been found to have roughly 1500 factories through immunofluorescence detection of RNAP II however cells taken from different tissues of the same mouse group had between 100 and 300 factories. Measurements of the number of transcription factories in HeLa cells give a varied result. For example, using the traditional fluorescence microscopy approach 300 – 500 factories were found but using both confocal and electron microscopy roughly 2100 were detected.

Factory Specialisation

In addition to the specialisation factories have for the type of RNA polymerase they contain, there is a further level of specialisation present. There are some factories that only transcribe a certain set of related genes, this further strengthens the concept that the main function of a transcription factory is for transcriptional efficiency.

Assembly and Maintenance

There is much debate to whether transcription factories assemble because of the transcriptional demands of the genome or if they are stable structures that are conserved over time. Experimentally, it appears that they remain fixed over a short period of time; newly made mRNA were pulse labelled over 15 minutes and it showed no new transcription factories appearing. This is also supported by inhibition experiments. In these studies heat shock was used to turn off transcription which resulted in no change in the number of polymerases detected. Upon further analysis of western blot data it was suggested that there was in fact a slight decrease over time of transcription factories. Therefore, it could be claimed that polymerase molecules are released gently over time from the factory when there is a lack of transcription which would eventually lead to the complete loss of the transcription factory.

There is also several pieces of evidence that promotes the idea of transcription factories assembling *de novo* due to transcriptional demands. GFP polymerase fluorescence experiments have shown that the inducement of transcription in *Drosophila* polytene nuclei leads to the formation of a factory which contradicts the notion of a stable and secure structure.

Mechanism

It was previously thought that it was the relatively small RNA polymerase that moves along the comparatively larger DNA template during transcription. However, increasing evidence supports the notion that due to the tethering of a transcription factory to the nuclear matrix, it is in fact the large DNA template that is moved to accommodate RNA polymerisation. *In vitro* studies for example have shown that RNA polymerases attached to a surface are capable of both rotating the DNA template and threading it through the polymerase to start transcription; which indicates the capabilities of RNA polymerase to be a molecular motor. Chromosome Conformation Capture (3C) also supports the idea of the DNA template diffusing towards a stationary RNA polymerase.

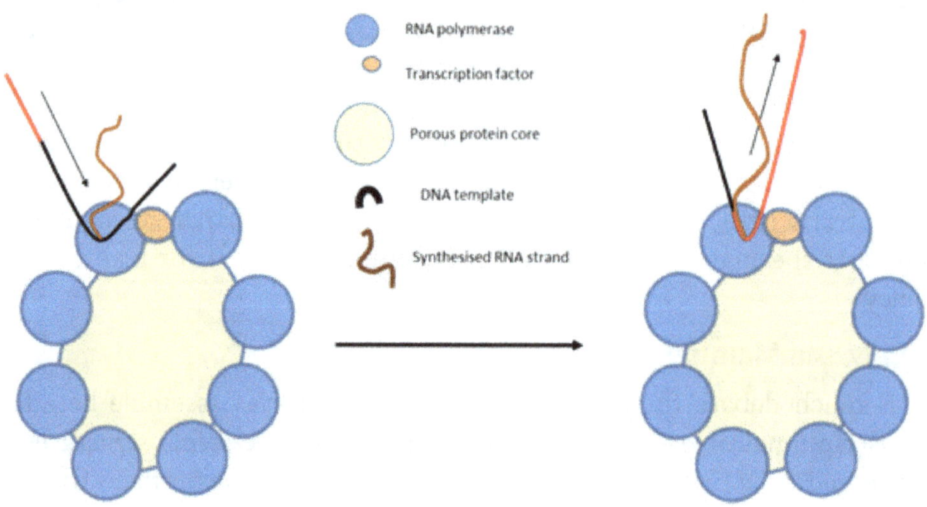

Fig: The Hypothesis that it is the Transcription Factory that Remains Immobilised during Transcription as opposed to the DNA Template. It shows how a Section of the Gene being Transcribed (Brown) gets Pulled and Shuttled through the RNA Polymerase during the Process

There remains a doubt to this mechanism of transcription. Firstly, it is unknown how a stationary polymerase is capable of transcribing genes on the (+)-strand and (-)-strand at the same genomic locus at the same time. This is in addition to a lack of conclusive evidence on how the polymerase remains immobilised (how it is tethered) and what structure it is tethered to.

Effect on Genomic and Nuclear Structure

There are several consequences the formation of a transcription factory has on nuclear and genomic structures. It has been proposed that the factories are responsible for nuclear organisation; they have been suggested to promote chromatin loop formation by two potential mechanisms:

The first mechanism suggests that loops form because 2 genes on the same chromosome require the same transcription machinery that would be found in a specific transcription factory. This requirement will attract the gene loci to the factory thus creating a loop.

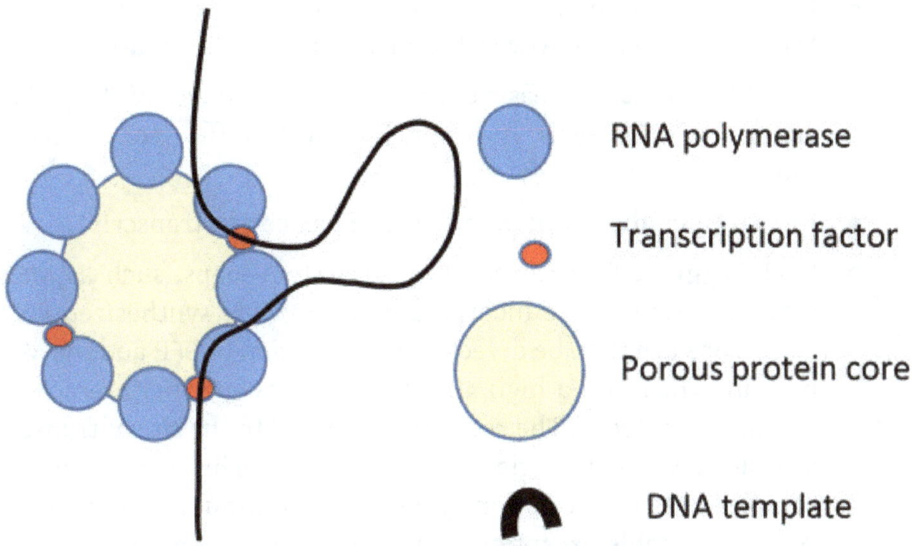

Fig: The Attraction of related Genes to RNAP and the Required Transcription Factors Causes the Formation of a Chromatin Loop, thereby Affecting the Genome Structure

Transcription factories are also suggested to be responsible for gene clustering, this is because related genes would require the same transcriptional machinery and if a factory satisfies these needs the genes would be attracted to the factory . While the clustering of genes can be beneficial for transcriptional efficiency, there could be negative consequences to this. Gene translocation events occur when genes are in close proximity to one another; which will occur more often when a transcriptional factory is present. Gene translocation events, like point mutations, generally are detrimental to the organism and so therefore could lead to the possibility of disease. However, on the other hand recent research has suggested that there is no correlation between inter-gene interactions and translocation frequencies.

2.3.6 Measuring and Detecting

Transcription can be measured and detected in a variety of ways:

- **G-Less Cassette Transcription Assay:** Measures promoter strength.
- **Run-Off Transcription Assay:** Identifies transcription start sites (TSS).
- **Nuclear Run-On Assay:** Measures the relative abundance of newly formed transcripts.
- **KAS-Seq:** Measures single-stranded DNA generated by RNA polymerases; can work with 1,000 cells.
- **RNase Protection Assay and ChIP-Chip of RNAP:** Detect active transcription sites.

- **RT-PCR:** Measures the absolute abundance of total or nuclear RNA levels, which may however differ from transcription rates.
- **DNA Microarrays:** Measures the relative abundance of the global total or nuclear RNA levels; however, these may differ from transcription rates.
- **In Situ Hybridization:** Detects the presence of a transcript.
- **MS2 Tagging:** By incorporating RNA stem loops, such as MS2, into a gene, these become incorporated into newly synthesized RNA. The stem loops can then be detected using a fusion of GFP and the MS2 coat protein, which has a high affinity, sequence-specific interaction with the MS2 stem loops. The recruitment of GFP to the site of transcription is visualized as a single fluorescent spot. This new approach has revealed that transcription occurs in discontinuous bursts, or pulses. With the notable exception of in situ techniques, most other methods provide cell population averages, and are not capable of detecting this fundamental property of genes.
- **Northern Blot:** The traditional method, and until the advent of RNA-Seq, the most quantitative.
- **RNA-Seq:** Applies next-generation sequencing techniques to sequence whole transcriptomes, which allows the measurement of relative abundance of RNA, as well as the detection of additional variations such as fusion genes, post-transcriptional edits and novel splice sites.
- **Single Cell RNA-Seq:** Amplifies and reads partial transcriptomes from isolated cells, allowing for detailed analyses of RNA in tissues, embryos, and cancers.

2.3.7 Reverse Transcription

Some viruses (such as HIV, the cause of AIDS), have the ability to transcribe RNA into DNA. HIV has an RNA genome that is *reverse transcribed* into DNA. The resulting DNA can be merged with the DNA genome of the host cell. The main enzyme responsible for synthesis of DNA from an RNA template is called reverse transcriptase.

In the case of HIV, reverse transcriptase is responsible for synthesizing a complementary DNA strand (cDNA) to the viral RNA genome. The enzyme ribonuclease H then digests the RNA strand, and reverse transcriptase synthesises a complementary strand of DNA to form a double helix DNA structure (cDNA). The cDNA is integrated into the host cell's genome by the enzyme integrase, which causes the host cell to generate viral proteins that reassemble into new viral particles. In HIV, subsequent to this, the host cell undergoes programmed cell death, or apoptosis, of T cells. However, in other retroviruses, the host cell remains intact as the virus buds out of the cell.

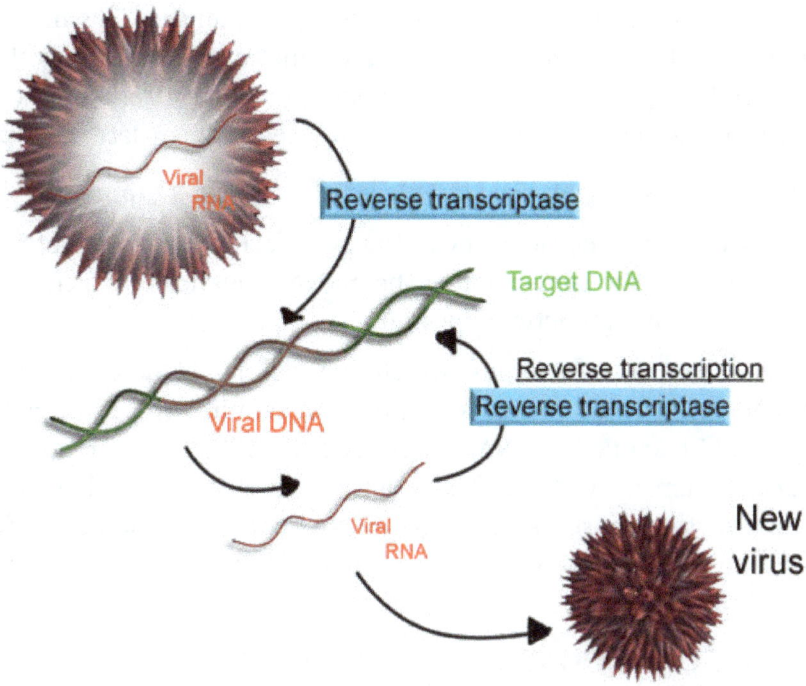

Fig: Scheme of Reverse Transcription

Some eukaryotic cells contain an enzyme with reverse transcription activity called telomerase. Telomerase carries an RNA template from which it synthesizes a telomere, a repeating sequence of DNA, to the end of linear chromosomes. It is important because every time a linear chromosome is duplicated, it is shortened. With the telomere at the ends of chromosomes, the shortening eliminates some of the non-essential, repeated sequence, rather than the protein-encoding DNA sequence farther away from the chromosome end.

Telomerase is often activated in cancer cells to enable cancer cells to duplicate their genomes indefinitely without losing important protein-coding DNA sequence. Activation of telomerase could be part of the process that allows cancer cells to become immortal. The immortalizing factor of cancer via telomere lengthening due to telomerase has been proven to occur in 90% of all carcinogenic tumors *in vivo* with the remaining 10% using an alternative telomere maintenance route called ALT or Alternative Lengthening of Telomeres.

2.4 RNA PROCESSING IN EUKARYOTIC CELLS

In eukaryotic cells, the primary RNA transcript undergoes several modifications before it becomes mature mRNA and exits the nucleus. This processing includes the addition of a 5' cap and a poly-A tail, which help protect the mRNA from degradation and facilitate its export from the nucleus to the cytoplasm. Additionally, introns—non-coding regions within the RNA transcript—are removed through a process called splicing, leaving only the exons, or coding sequences. This mature mRNA sequence, now containing only the necessary genetic information, is then ready for translation. RNA processing is a critical step in gene expression in eukaryotic organisms, ensuring that only the correct sequences are translated into proteins.

2.5 TRANSLATION: CONVERTING MRNA TO PROTEIN

Once the mRNA reaches the ribosome in the cytoplasm, the process of translation begins. Translation is the conversion of the mRNA sequence into a chain of amino acids that will fold into a functional protein.

This process occurs on ribosomes, which are cellular structures composed of ribosomal RNA (rRNA) and proteins. Translation requires the coordinated effort of three types of RNA: mRNA, rRNA, and transfer RNA (tRNA). Each tRNA molecule carries a specific amino acid and has an anticodon that is complementary to an mRNA codon. As the ribosome reads the mRNA sequence codon by codon, tRNA molecules bring the corresponding amino acids, which are then linked together in a growing polypeptide chain.

2.6 THE ROLE OF CODONS IN TRANSLATION

During translation, the ribosome reads the mRNA sequence in groups of three nucleotides called codons. Each codon specifies a particular amino acid, and the sequence of codons in the mRNA determines the order of amino acids in the resulting protein. The genetic code, a set of rules by which codons correspond to specific amino acids, is nearly universal across all living organisms, underscoring the fundamental role of DNA in life. Some codons serve as "start" or "stop" signals, indicating where translation should begin and end. The start codon (AUG) also codes for the amino acid methionine, which initiates the polypeptide chain, while stop codons (UAA, UAG, UGA) signal the termination of translation.

2.7 PROTEIN FOLDING AND POST-TRANSLATIONAL MODIFICATIONS

After the polypeptide chain is synthesized, it must fold into a specific three-dimensional structure to become a functional protein. This folding process is guided by the amino acid sequence and is essential for the protein's biological activity. In some cases, chaperone proteins assist in folding to ensure the protein achieves the correct shape. Additionally, many proteins undergo post-translational modifications, such as phosphorylation, glycosylation, or cleavage, which further refine the protein's function and stability. These modifications can alter the protein's activity, localization, or interaction with other cellular molecules, making them critical for the protein's ultimate role in cellular processes.

2.8 REGULATION OF GENE EXPRESSION

The flow of information from DNA to protein is tightly regulated at multiple levels to ensure that proteins are produced at the right time, in the right place, and in the correct amounts. Gene expression can be controlled during transcription, mRNA processing, translation, and even post-translation. Regulatory proteins, signaling molecules, and environmental factors can influence gene expression. For example, transcription factors are proteins that bind to specific DNA sequences to either promote or inhibit transcription, controlling which genes are expressed in a given cell. This regulation is essential for cellular differentiation, adaptation to environmental changes, and maintaining cellular homeostasis.

2.9 MUTATIONS AND THEIR IMPACT ON INFORMATION FLOW

Mutations, or changes in the DNA sequence, can disrupt the flow of information from DNA to protein and lead to altered or non-functional proteins. Point mutations, insertions, deletions, and frameshift mutations can affect the mRNA sequence, potentially leading to incorrect amino acid sequences or premature stop codons in proteins.

Some mutations are silent, meaning they do not alter the resulting protein, while others can have significant effects, leading to diseases or cellular dysfunction. Understanding how mutations affect protein synthesis is crucial in fields like genetics, medicine, and evolutionary biology, where researchers investigate how genetic changes influence phenotype and inheritance.

2.10 ADVANCES IN UNDERSTANDING THE DNA-PROTEIN PATHWAY

Modern research has expanded our understanding of the flow of information from DNA to protein. Techniques like polymerase chain reaction (PCR), gene sequencing, and CRISPR gene editing have allowed scientists to manipulate and study genes with unprecedented precision. These tools have improved our ability to explore how genetic variations affect protein function and how certain genetic sequences are regulated.

Additionally, proteomics—the study of the full set of proteins in a cell or organism—provides insights into protein function, interactions, and regulation, deepening our knowledge of cellular mechanisms. These advancements highlight the importance of the DNA-to-protein pathway in understanding biological systems and developing new therapies for genetic disorders.

2.11 CONCLUSION

The flow of information from DNA to protein is fundamental to the survival and function of all living organisms. From transcription and translation to protein folding and regulation, each step in this process is finely tuned to ensure that genetic information is accurately expressed.

This pathway enables cells to carry out complex functions, adapt to environmental changes, and maintain homeostasis. The central dogma not only provides a framework for understanding genetics and molecular biology but also has practical applications in medicine, biotechnology, and evolutionary studies. As research continues to unravel the intricacies of the DNA-to-protein pathway, our ability to manipulate and harness this process holds promising potential for advancing science and improving human health.

CHAPTER 3
CONTROL OF GENE EXPRESSION

3.1 INTRODUCTION TO GENE EXPRESSION

Gene expression is the process by which information from a gene is used in the synthesis of a functional gene product that enables it to produce end products, proteins or non-coding RNA, and ultimately affect a phenotype. These products are often proteins, but in non-protein-coding genes such as transfer RNA (tRNA) and small nuclear RNA (snRNA), the product is a functional non-coding RNA. The process of gene expression is used by all known life—eukaryotes (including multicellular organisms), prokaryotes (bacteria and archaea), and utilized by viruses—to generate the macromolecular machinery for life.

In genetics, gene expression is the most fundamental level at which the genotype gives rise to the phenotype, *i.e.*, observable trait. The genetic information stored in DNA represents the genotype, whereas the phenotype results from the "interpretation" of that information. Such phenotypes are often displayed by the synthesis of proteins that control the organism's structure and development, or that act as enzymes catalyzing specific metabolic pathways.

All steps in the gene expression process may be modulated (regulated), including the transcription, RNA splicing, translation, and post-translational modification of a protein. Regulation of gene expression gives control over the timing, location, and amount of a given gene product (protein or ncRNA) present in a cell and can have a profound effect on the cellular structure and function. Regulation of gene expression is the basis for cellular differentiation, development, morphogenesis and the versatility and adaptability of any organism. Gene regulation may therefore serve as a substrate for evolutionary change.

3.1.1 Mechanism

Transcription

The production of a RNA copy from a DNA strand is called transcription, and is performed by RNA polymerases, which add one ribonucleotide at a time to a growing RNA strand as per the complementarity law of the nucleotide bases. This RNA is complementary to the template 3' → 5' DNA strand, with the exception that thymines (T) are replaced with uracils (U) in the RNA and possible errors.

In bacteria, transcription is carried out by a single type of RNA polymerase, which needs to bind a DNA sequence called a Pribnow box with the help of the sigma factor protein (σ factor) to start transcription. In eukaryotes, transcription is performed in the nucleus by three types of RNA polymerases, each of which needs a special DNA sequence called the promoter and a set of DNA-binding proteins—transcription factors—to initiate the process. RNA polymerase I is responsible for transcription of ribosomal RNA (rRNA) genes. RNA polymerase II (Pol II) transcribes all protein-coding genes but also some non-coding RNAs (*e.g.*, snRNAs, snoRNAs or long non-coding RNAs). RNA polymerase III transcribes 5S rRNA, transfer RNA (tRNA) genes, and some small non-coding RNAs (*e.g.*, 7SK). Transcription ends when the polymerase encounters a sequence called the terminator.

mRNA Processing

While transcription of prokaryotic protein-coding genes creates messenger RNA (mRNA) that is ready for translation into protein, transcription of eukaryotic genes leaves a primary transcript of RNA (pre-RNA), which first has to undergo a series of modifications to become a mature RNA.

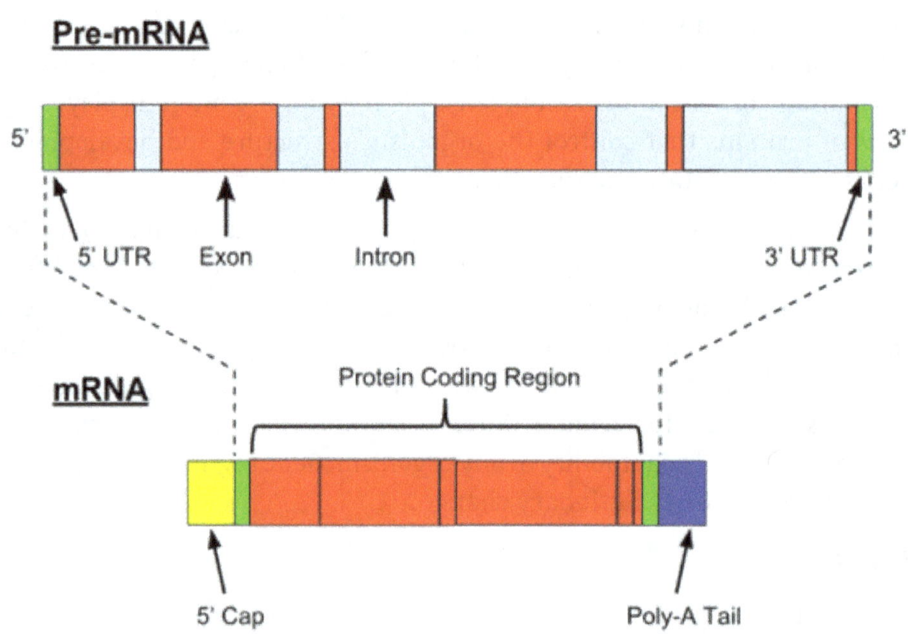

Fig: Illustration of Exons and Introns in Pre-mRNA and the Formation of Mature mRNA by Splicing. The UTRs (in Green) are Non-Coding Parts of Exons at the Ends of the mRNA

Types and steps involved in the maturation processes vary between coding and non-coding preRNAs; *i.e.*, even though preRNA molecules for both mRNA and tRNA undergo splicing, the steps and machinery involved are different. The processing of pre-mRNA include 5' *capping*, which is set of enzymatic reactions that add 7-methylguanosine (m^7G) to the 5' end of pre-mRNA and thus protect the RNA from degradation by exonucleases. The m^7G cap is then bound by cap binding complex heterodimer (CBC20/CBC80), which aids in mRNA export to cytoplasm and also protect the RNA from decapping.

Another modification is 3' *cleavage and polyadenylation*. They occur if polyadenylation signal sequence (5'- AAUAAA-3') is present in pre-mRNA, which is usually between protein-coding sequence and terminator. The pre-mRNA is first cleaved and then a series of ~200 adenines (A) are added to form poly(A) tail, which protects the RNA from degradation. The poly(A) tail is bound by multiple poly(A)-binding proteins (PABPs) necessary for mRNA export and translation re-initiation. In the inverse process of deadenylation, poly(A) tails are shortened by the CCR4-Not 3'-5' exonuclease, which often leads to full transcript decay.

A very important modification of eukaryotic pre-mRNA is *RNA splicing*. The majority of eukaryotic pre-mRNAs consist of alternating segments called exons and introns. During the process of splicing, an RNA-protein catalytical complex known as spliceosome catalyzes two transesterification reactions, which remove an intron and release it in form of lariat structure, and then splice neighbouring exons together. In certain cases, some introns or exons can be either removed or retained in mature mRNA. This so-called alternative splicing creates series of different transcripts originating from a single gene. Because these transcripts can be potentially translated into different proteins, splicing extends the complexity of eukaryotic gene expression and the size of a species proteome.

Extensive RNA processing may be an evolutionary advantage made possible by the nucleus of eukaryotes. In prokaryotes, transcription and translation happen together, whilst in eukaryotes, the nuclear membrane separates the two processes, giving time for RNA processing to occur.

Non-Coding RNA Maturation

In most organisms non-coding genes (ncRNA) are transcribed as precursors that undergo further processing. In the case of ribosomal RNAs (rRNA), they are often transcribed as a pre-rRNA that contains one or more rRNAs. The pre-rRNA is cleaved and modified (2'-*O*-methylation and pseudouridine formation) at specific sites by approximately 150 different small nucleolus-restricted RNA species, called snoRNAs. SnoRNAs associate with proteins, forming snoRNPs. While snoRNA part basepair with the target RNA and thus position the modification at a precise site, the protein part performs the catalytical reaction. In eukaryotes, in particular a snoRNP called RNase, MRP cleaves the 45S pre-rRNA into the 28S, 5.8S, and 18S rRNAs. The rRNA and RNA processing factors form large aggregates called the nucleolus.

In the case of transfer RNA (tRNA), for example, the 5' sequence is removed by RNase P, whereas the 3' end is removed by the tRNase Z enzyme and the non-templated 3' CCA tail is added by a nucleotidyl transferase. In the case of micro RNA (miRNA), miRNAs are first transcribed as primary transcripts or pri-miRNA with a cap and poly-A tail and processed to short, 70-nucleotide stem-loop structures known as pre-miRNA in the cell nucleus by the enzymes Drosha and Pasha. After being exported, it is then processed to mature miRNAs in the cytoplasm by interaction with the endonuclease Dicer, which also initiates the formation of the RNA-induced silencing complex (RISC), composed of the Argonaute protein.

Even snRNAs and snoRNAs themselves undergo series of modification before they become part of functional RNP complex. This is done either in the nucleoplasm or in the specialized compartments called Cajal bodies. Their bases are methylated or pseudouridinilated by a group of small Cajal body-specific RNAs (scaRNAs), which are structurally similar to snoRNAs.

RNA Export

In eukaryotes most mature RNA must be exported to the cytoplasm from the nucleus. While some RNAs function in the nucleus, many RNAs are transported through the nuclear pores and into the cytosol. Export of RNAs requires association with specific proteins known as exportins. Specific exportin molecules are responsible for the export of a given RNA type. mRNA transport also requires the correct association with Exon Junction Complex (EJC), which ensures that correct processing of the mRNA is completed before export. In some cases RNAs are additionally transported to a specific part of the cytoplasm, such as a synapse; they are then towed by motor proteins that bind through linker proteins to specific sequences (called "zipcodes") on the RNA.

Translation

For some non-coding RNA, the mature RNA is the final gene product. In the case of messenger RNA (mRNA) the RNA is an information carrier coding for the synthesis of one or more proteins. mRNA carrying a single protein sequence (common in eukaryotes) is monocistronic whilst mRNA carrying multiple protein sequences (common in prokaryotes) is known as polycistronic.

Every mRNA consists of three parts: a 5' untranslated region (5'UTR), a protein-coding region or open reading frame (ORF), and a 3' untranslated region (3'UTR). The coding region carries information for protein synthesis encoded by the genetic code to form triplets. Each triplet of nucleotides of the coding region is called a codon and corresponds to a binding site complementary to an anticodon triplet in transfer RNA. Transfer RNAs with the same anticodon sequence always carry an identical type of amino acid. Amino acids are then chained together by the ribosome according to the order of triplets in the coding region. The ribosome helps transfer RNA to bind to messenger RNA and takes the amino acid from each transfer RNA and makes a structure-less protein out of it. Each mRNA molecule is translated into many protein molecules, on average ~2800 in mammals.

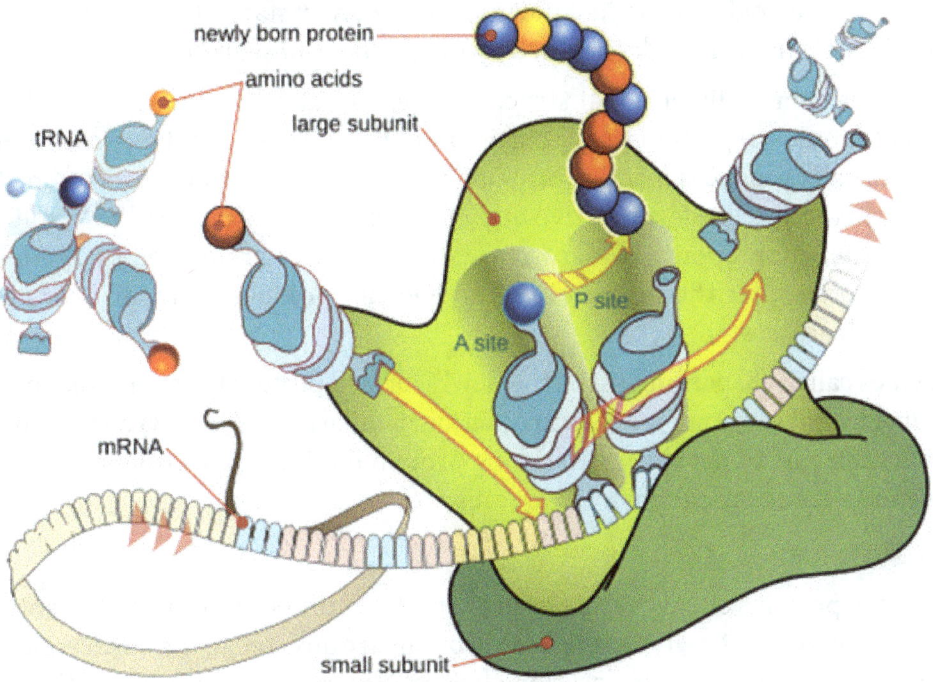

Fig: During the Translation, tRNA Charged with Amino Acid Enters the Ribosome and Aligns with the Correct mRNA Triplet. Ribosome then Adds Amino Acid to Growing Protein Chain

In prokaryotes translation generally occurs at the point of transcription (co-transcriptionally), often using a messenger RNA that is still in the process of being created. In eukaryotes translation can occur in a variety of regions of the cell depending on where the protein being written is supposed to be. Major locations are the cytoplasm for soluble cytoplasmic proteins and the membrane of the endoplasmic reticulum for proteins that are for export from the cell or insertion into a cell membrane.

Proteins that are supposed to be produced at the endoplasmic reticulum are recognised part-way through the translation process. This is governed by the signal recognition particle—a protein that binds to the ribosome and directs it to the endoplasmic reticulum when it finds a signal peptide on the growing (nascent) amino acid chain.

Folding

Each protein exists as an unfolded polypeptide or random coil when translated from a sequence of mRNA into a linear chain of amino acids. This polypeptide lacks any developed three-dimensional structure. The polypeptide then folds into its characteristic and functional three-dimensional structure from a random coil. Amino acids interact with each other to produce a well-defined three-dimensional structure, the folded protein known as the native state. The resulting three-dimensional structure is determined by the amino acid sequence.

The correct three-dimensional structure is essential to function, although some parts of functional proteins may remain unfolded. Failure to fold into the intended shape usually produces inactive proteins with different properties including toxic prions. Several neurodegenerative and other diseases are believed to result from the accumulation of *misfolded* proteins. Many allergies are caused by the folding of the proteins, for the immune system does not produce antibodies for certain protein structures.

Enzymes called chaperones assist the newly formed protein to attain (fold into) the 3-dimensional structure it needs to function. Similarly, RNA chaperones help RNAs attain their functional shapes. Assisting protein folding is one of the main roles of the endoplasmic reticulum in eukaryotes.

Translocation

Secretory proteins of eukaryotes or prokaryotes must be translocated to enter the secretory pathway. Newly synthesized proteins are directed to the eukaryotic Sec61 or prokaryotic SecYEG translocation channel by signal peptides. The efficiency of protein secretion in eukaryotes is very dependent on the signal peptide which has been used.

Protein Transport

Many proteins are destined for other parts of the cell than the cytosol and a wide range of signalling sequences or (signal peptides) are used to direct proteins to where they are supposed to be. In prokaryotes this is normally a simple process due to limited compartmentalisation of the cell. However, in eukaryotes there is a great variety of different targeting processes to ensure the protein arrives at the correct organelle.

Not all proteins remain within the cell and many are exported, for example, digestive enzymes, hormones and extracellular matrix proteins. In eukaryotes the export pathway is well developed and the main mechanism for the export of these proteins is translocation to the endoplasmic reticulum, followed by transport via the Golgi apparatus.

3.1.2 Regulation of Gene Expression

Regulation of gene expression is the control of the amount and timing of appearance of the functional product of a gene. Control of expression is vital to allow a cell to produce the gene products it needs when it needs them; in turn, this gives cells the flexibility to adapt to a variable environment, external signals, damage to the cell, and other stimuli. More generally, gene regulation gives the cell control over all structure and function, and is the basis for cellular differentiation, morphogenesis and the versatility and adaptability of any organism.

Numerous terms are used to describe types of genes depending on how they are regulated; these include:

- A constitutive gene is a gene that is transcribed continually as opposed to a facultative gene, which is only transcribed when needed.
- A *housekeeping gene* is a gene that is required to maintain basic cellular function and so is typically expressed in all cell types of an organism. Examples include actin, GAPDH and ubiquitin. Some housekeeping genes are transcribed at a relatively constant rate and these genes can be used as a reference point in experiments to measure the expression rates of other genes.
- A facultative gene is a gene only transcribed when needed as opposed to a constitutive gene.
- An inducible gene is a gene whose expression is either responsive to environmental change or dependent on the position in the cell cycle.

Any step of gene expression may be modulated, from the DNA-RNA transcription step to post-translational modification of a protein. The stability of the final gene product, whether it is RNA or protein, also contributes to the expression level of the gene—an unstable product results in a low expression level. In general gene expression is regulated through changes in the number and type of interactions between molecules that collectively influence transcription of DNA and translation of RNA.

Some simple examples of where gene expression is important are:

- Control of insulin expression so it gives a signal for blood glucose regulation.
- X chromosome inactivation in female mammals to prevent an "overdose" of the genes it contains.
- Cyclin expression levels control progression through the eukaryotic cell cycle.

Transcriptional Regulation in Learning and Memory

In a rat, contextual fear conditioning (CFC) is a painful learning experience. Just one episode of CFC can result in a life-long fearful memory. After an episode of CFC, cytosine methylation is altered in the promoter regions of about 9.17% of all genes in the hippocampus neuron DNA of a rat. The hippocampus is where new memories are initially stored.

After CFC about 500 genes have increased transcription (often due to demethylation of CpG sites in a promoter region) and about 1,000 genes have decreased transcription (often due to newly formed 5-methylcytosine at CpG sites in a promoter region). The pattern of induced and repressed genes within neurons appears to provide a molecular basis for forming the first transient memory of this training event in the hippocampus of the rat brain.

Some specific mechanisms guiding new DNA methylations and new DNA demethylations in the hippocampus during memory establishment have been established. One mechanism includes guiding the short isoform of the TET1 DNA demethylation enzyme, TET1s, to about 600 locations on the genome. The guidance is performed by association of TET1s with EGR1 protein, a transcription factor important in memory formation. Bringing TET1s to these locations initiates DNA demethylation at those sites, up-regulating associated genes. A second mechanism involves DNMT3A2, a splice-isoform of DNA methyltransferase DNMT3A, which adds methyl groups to cytosines in DNA. This isoform is induced by synaptic activity, and its location of action appears to be determined by histone post-translational modifications (a histone code). The resulting new messenger RNAs are then transported by messenger RNP particles (neuronal granules) to synapses of the neurons, where they can be translated into proteins affecting the activities of synapses.

In particular, the brain-derived neurotrophic factor gene (*BDNF*) is known as a "learning gene". After CFC there was upregulation of *BDNF* gene expression, related to decreased CpG methylation of certain internal promoters of the gene, and this was correlated with learning.

Transcriptional Regulation in Cancer

The majority of gene promoters contain a CpG island with numerous CpG sites. When many of a gene's promoter CpG sites are methylated the gene becomes silenced. Colorectal cancers typically have 3 to 6 driver mutations and 33 to 66 hitchhiker or passenger mutations. However, transcriptional silencing may be of more importance than mutation in causing progression to cancer. For example, in colorectal cancers about 600 to 800 genes are transcriptionally silenced by CpG island methylation. Transcriptional repression in cancer can also occur by other epigenetic mechanisms, such as altered expression of microRNAs. In breast cancer, transcriptional repression of BRCA1 may occur more frequently by over-transcribed microRNA-182 than by hypermethylation of the BRCA1 promoter.

Post-Transcriptional Regulation

In eukaryotes, where export of RNA is required before translation is possible, nuclear export is thought to provide additional control over gene expression. All transport in and out of the nucleus is via the nuclear pore and transport is controlled by a wide range of importin and exportin proteins.

Expression of a gene coding for a protein is only possible if the messenger RNA carrying the code survives long enough to be translated. In a typical cell, an RNA molecule is only stable if specifically protected from degradation. RNA degradation has particular importance in regulation of expression in eukaryotic cells where mRNA has to travel significant distances before being translated. In eukaryotes, RNA is stabilised by certain post-transcriptional modifications, particularly the 5' cap and poly-adenylated tail.

Intentional degradation of mRNA is used not just as a defence mechanism from foreign RNA (normally from viruses) but also as a route of mRNA *destabilisation*. If an mRNA molecule has a complementary sequence to a small interfering RNA then it is targeted for destruction via the RNA interference pathway.

Three Prime Untranslated Regions and MicroRNAs

Three prime untranslated regions (3'UTRs) of messenger RNAs (mRNAs) often contain regulatory sequences that post-transcriptionally influence gene expression. Such 3'-UTRs often contain both binding sites for microRNAs (miRNAs) as well as for regulatory proteins. By binding to specific sites within the 3'-UTR, miRNAs can decrease gene expression of various mRNAs by either inhibiting translation or directly causing degradation of the transcript. The 3'-UTR also may have silencer regions that bind repressor proteins that inhibit the expression of a mRNA.

The 3'-UTR often contains microRNA response elements (MREs). MREs are sequences to which miRNAs bind. These are prevalent motifs within 3'-UTRs. Among all regulatory motifs within the 3'-UTRs (e.g., including silencer regions), MREs make up about half of the motifs.

As of 2014, the miRBase web site, an archive of miRNA sequences and annotations, listed 28,645 entries in 233 biologic species. Of these, 1,881 miRNAs were in annotated human miRNA loci. miRNAs were predicted to have an average of about four hundred target mRNAs (affecting expression of several hundred genes). Friedman et al. estimate that >45,000 miRNA target sites within human mRNA 3'UTRs are conserved above background levels, and >60% of human protein-coding genes have been under selective pressure to maintain pairing to miRNAs. Direct experiments show that a single miRNA can reduce the stability of hundreds of unique mRNAs. Other experiments show that a single miRNA may repress the production of hundreds of proteins, but that this repression often is relatively mild (less than 2-fold).

The effects of miRNA dysregulation of gene expression seem to be important in cancer. For instance, in gastrointestinal cancers, nine miRNAs have been identified as epigenetically altered and effective in down regulating DNA repair enzymes. The effects of miRNA dysregulation of gene expression also seem to be important in neuropsychiatric disorders, such as schizophrenia, bipolar disorder, major depression, Parkinson's disease, Alzheimer's disease and autism spectrum disorders.

Translational Regulation

Direct regulation of translation is less prevalent than control of transcription or mRNA stability but is occasionally used. Inhibition of protein translation is a major target for toxins and antibiotics, so they can kill a cell by overriding its normal gene expression control. Protein synthesis inhibitors include the antibiotic neomycin and the toxin ricin.

Post-Translational Modifications

Post-translational modifications (PTMs) are covalent modifications to proteins. Like RNA splicing, they help to significantly diversify the proteome. These modifications are usually catalyzed by enzymes. Additionally, processes like covalent additions to amino acid side chain residues can often be reversed by other enzymes. However, some, like the proteolytic cleavage of the protein backbone, are irreversible.

PTMs play many important roles in the cell. For example, phosphorylation is primarily involved in activating and deactivating proteins and in signaling pathways. PTMs are involved in transcriptional regulation: an important function of acetylation and methylation is histone tail modification, which alters how accessible DNA is for transcription. They can also be seen in the immune system, where glycosylation plays a key role. One type of PTM can initiate another type of PTM. Proteolysis, other than being involved in breaking down proteins, is also important in activating and deactivating them, and in regulating biological processes such as DNA transcription and cell death.

3.1.3 Measurement

Measuring gene expression is an important part of many life sciences, as the ability to quantify the level at which a particular gene is expressed within a cell, tissue or organism can provide a lot of valuable information. For example, measuring gene expression can:

- Identify viral infection of a cell (viral protein expression).
- Determine an individual's susceptibility to cancer (oncogene expression).
- Find if a bacterium is resistant to penicillin (beta-lactamase expression).
- Gene expression profiling evaluates a panel of genes to help understand the fundamental mechanism of a cell. This is increasingly used in cancer therapy to target specific chemotherapy.

Similarly, the analysis of the location of protein expression is a powerful tool, and this can be done on an organismal or cellular scale. Investigation of localization is particularly important for the study of development in multicellular organisms and as an indicator of protein function in single cells. Ideally, measurement of expression is done by detecting the final gene product (for many genes, this is the protein); however, it is often easier to detect one of the precursors, typically mRNA and to infer gene-expression levels from these measurements.

mRNA Quantification

Levels of mRNA can be quantitatively measured by northern blotting, which provides size and sequence information about the mRNA molecules. A sample of RNA is separated on an agarose gel and hybridized to a radioactively labeled RNA probe that is complementary to the target sequence. The radiolabeled RNA is then detected by an autoradiograph. Because the use of radioactive reagents makes the procedure time-consuming and potentially dangerous, alternative labeling and detection methods, such as digoxigenin and biotin chemistries, have been developed. Perceived disadvantages of Northern blotting are that large quantities of RNA are required and that quantification may not be completely accurate, as it involves measuring band strength in an image of a gel. On the other hand, the additional mRNA size information from the Northern blot allows the discrimination of alternately spliced transcripts.

Another approach for measuring mRNA abundance is RT-qPCR. In this technique, reverse transcription is followed by quantitative PCR. Reverse transcription first generates a DNA template from the mRNA; this single-stranded template is called cDNA.

The cDNA template is then amplified in the quantitative step, during which the fluorescence emitted by labeled hybridization probes or intercalating dyes changes as the DNA amplification process progresses. With a carefully constructed standard curve, qPCR can produce an absolute measurement of the number of copies of original mRNA, typically in units of copies per nanolitre of homogenized tissue or copies per cell. qPCR is very sensitive (detection of a single mRNA molecule is theoretically possible), but can be expensive depending on the type of reporter used; fluorescently labeled oligonucleotide probes are more expensive than non-specific intercalating fluorescent dyes.

For expression profiling, or high-throughput analysis of many genes within a sample, quantitative PCR may be performed for hundreds of genes simultaneously in the case of low-density arrays. A second approach is the hybridization microarray. A single array or "chip" may contain probes to determine transcript levels for every known gene in the genome of one or more organisms. Alternatively, "tag based" technologies like Serial analysis of gene expression (SAGE) and RNA-Seq, which can provide a relative measure of the cellular concentration of different mRNAs, can be used. An advantage of tag-based methods is the "open architecture", allowing for the exact measurement of any transcript, with a known or unknown sequence. Next-generation sequencing (NGS) such as RNA-Seq is another approach, producing vast quantities of sequence data that can be matched to a reference genome. Although NGS is comparatively time-consuming, expensive, and resource-intensive, it can identify single-nucleotide polymorphisms, splice-variants, and novel genes, and can also be used to profile expression in organisms for which little or no sequence information is available.

Protein Quantification

For genes encoding proteins, the expression level can be directly assessed by a number of methods with some clear analogies to the techniques for mRNA quantification. One of the most commonly used methods is to perform a Western blot against the protein of interest. This gives information on the size of the protein in addition to its identity. A sample (often cellular lysate) is separated on a polyacrylamide gel, transferred to a membrane and then probed with an antibody to the protein of interest.

The antibody can either be conjugated to a fluorophore or to horseradish peroxidase for imaging and/or quantification. The gel-based nature of this assay makes quantification less accurate, but it has the advantage of being able to identify later modifications to the protein, for example proteolysis or ubiquitination, from changes in size.

mRNA-Protein Correlation

While transcription directly reflects gene expression, the copy number of mRNA molecules does not directly correlate with the number of protein molecules translated from mRNA. Quantification of both protein and mRNA permits a correlation of the two levels. Regulation on each step of gene expression can impact the correlation, as shown for regulation of translation or protein stability. Post-translational factors, such as protein transport in highly polar cells, can influence the measured mRNA-protein correlation as well.

Localization

Analysis of expression is not limited to quantification; localization can also be determined. mRNA can be detected with a suitably labelled complementary mRNA strand and protein can be detected via labelled antibodies. The probed sample is then observed by microscopy to identify where the mRNA or protein is.

By replacing the gene with a new version fused to a green fluorescent protein marker or similar, expression may be directly quantified in live cells. This is done by imaging using a fluorescence microscope. It is very difficult to clone a GFP-fused protein into its native location in the genome without affecting expression levels, so this method often cannot be used to measure endogenous gene expression. It is, however, widely used to measure the expression of a gene artificially introduced into the cell, for example via an expression vector. By fusing a target protein to a fluorescent reporter, the protein's behavior, including its cellular localization and expression level, can be significantly changed.

The enzyme-linked immunosorbent assay works by using antibodies immobilised on a microtiter plate to capture proteins of interest from samples added to the well. Using a detection antibody conjugated to an enzyme or fluorophore the quantity of bound protein can be accurately measured by fluorometric or colourimetric detection. The detection process is very similar to that of a Western blot, but by avoiding the gel steps more accurate quantification can be achieved.

3.1.4 Expression System

An expression system is a system specifically designed for the production of a gene product of choice. This is normally a protein although may also be RNA, such as tRNA or a ribozyme. An expression system consists of a gene, normally encoded by DNA, and the molecular machinery required to transcribe the DNA into mRNA and translate the mRNA into protein using the reagents provided. In the broadest sense this includes every living cell but the term is more normally used to refer to expression as a laboratory tool. An expression system is therefore often artificial in some manner. Expression systems are, however, a fundamentally natural process. Viruses are an excellent example where they replicate by using the host cell as an expression system for the viral proteins and genome.

- **Inducible Expression:** Doxycycline is also used in "Tet-on" and "Tet-off" tetracycline controlled transcriptional activation to regulate transgene expression in organisms and cell cultures.
- **In Nature:** In addition to these biological tools, certain naturally observed configurations of DNA (genes, promoters, enhancers, repressors) and the associated machinery itself are referred to as an expression system. This term is normally used in the case where a gene

or set of genes is switched on under well defined conditions, for example, the simple repressor switch expression system in Lambda phage and the lac operator system in bacteria. Several natural expression systems are directly used or modified and used for artificial expression systems such as the Tet-on and Tet-off expression system.

3.1.5 Gene Networks

Genes have sometimes been regarded as nodes in a network, with inputs being proteins such as transcription factors, and outputs being the level of gene expression. The node itself performs a function, and the operation of these functions have been interpreted as performing a kind of information processing within cells and determines cellular behavior.

Gene networks can also be constructed without formulating an explicit causal model. This is often the case when assembling networks from large expression data sets. Covariation and correlation of expression is computed across a large sample of cases and measurements (often transcriptome or proteome data). The source of variation can be either experimental or natural (observational). There are several ways to construct gene expression networks, but one common approach is to compute a matrix of all pair-wise correlations of expression across conditions, time points, or individuals and convert the matrix (after thresholding at some cut-off value) into a graphical representation in which nodes represent genes, transcripts, or proteins and edges connecting these nodes represent the strength of association.

3.2 LEVELS OF GENE EXPRESSION CONTROL

Gene expression can be controlled at multiple levels, from the DNA itself to the final protein product. The primary levels of gene regulation include transcriptional, post-transcriptional, translational, and post-translational control. At each of these levels, various molecular mechanisms and regulatory molecules interact to fine-tune the expression of specific genes, ensuring that cellular processes remain balanced. Different mechanisms are used depending on the organism, cell type, and environmental conditions, reflecting the complex and adaptable nature of gene regulation.

3.3 TRANSCRIPTIONAL CONTROL: THE FIRST STEP IN GENE REGULATION

Transcriptional control is the regulation of the process by which DNA is transcribed into messenger RNA (mRNA). It is one of the most efficient and commonly used methods of gene regulation, as it prevents unnecessary synthesis of mRNA and downstream products. Transcriptional control is mainly achieved by regulatory proteins called transcription factors, which bind to specific DNA sequences called promoters and enhancers. These proteins either activate or repress the transcription of the associated genes, depending on the cell's needs. For example, activator proteins can enhance gene expression by helping RNA polymerase bind to the promoter, while repressor proteins inhibit transcription by blocking RNA polymerase.

In addition to transcription factors, chromatin structure plays a significant role in transcriptional control. DNA is packaged into chromatin, a structure of DNA and histone proteins, which can be tightly packed (heterochromatin) or loosely packed (euchromatin). Genes within tightly packed chromatin are often inaccessible to transcription machinery and thus are silenced. In contrast, genes in loosely packed chromatin are more readily transcribed. Chromatin remodeling enzymes, such as histone acetyltransferases and deacetylases, modify histones to either relax or condense chromatin, providing a dynamic and reversible means of transcriptional regulation.

3.4 REGULATION

Epigenetic regulation refers to heritable changes in gene expression that do not involve changes to the DNA sequence. Epigenetic modifications include DNA methylation, histone modification, and non-coding RNA interactions. DNA methylation, the addition of a methyl group to cytosine bases, typically silences gene expression. This process is crucial in processes like cell differentiation, where specific genes are permanently turned off in certain cell types.

Histone modifications also contribute to epigenetic control by altering chromatin structure and influencing gene accessibility. These modifications can have lasting effects on gene expression patterns, especially in response to environmental factors. Epigenetic regulation provides a flexible, adaptive method for controlling gene expression and is essential for development, cellular memory, and response to environmental changes.

3.5 POST-TRANSCRIPTIONAL CONTROL: REGULATION OF MRNA PROCESSING AND STABILITY

Post-transcriptional control of gene expression occurs after mRNA is synthesized but before it is translated into protein. This level of control includes mRNA splicing, editing, transport, stability, and degradation. In eukaryotic cells, mRNA splicing is a critical step where introns (non-coding regions) are removed, and exons (coding regions) are joined. Alternative splicing, where different exons are included or excluded, allows for the production of multiple protein variants from a single gene, increasing protein diversity and enabling cells to adapt to different conditions.

mRNA stability is another significant aspect of post-transcriptional regulation. The lifespan of an mRNA molecule influences the amount of protein it can produce. Specific sequences in the mRNA, such as AU-rich elements, can make mRNA more susceptible to degradation, while regulatory proteins and microRNAs (miRNAs) can bind to mRNA and either stabilize or destabilize it.

miRNAs are small, non-coding RNAs that bind to complementary sequences in target mRNA molecules, often leading to their degradation or inhibition of translation. This mechanism allows cells to fine-tune protein production rapidly in response to cellular signals.

3.6 TRANSLATIONAL CONTROL: REGULATION AT THE RIBOSOME

Translational control regulates the process by which mRNA is translated into a polypeptide chain. This level of regulation is crucial in situations where a rapid response is needed, as it allows for the immediate increase or decrease of protein production without the need for new mRNA synthesis.

Certain proteins and non-coding RNAs bind to mRNA molecules to influence their translation at the ribosome. For example, translational repressors can bind to the 5' untranslated region (UTR) of mRNA to prevent the ribosome from initiating translation.

Global translational control mechanisms also exist and often come into play under stress conditions, such as nutrient deprivation or heat shock. During such times, cells may downregulate overall protein synthesis to conserve resources, while selectively translating stress-response proteins that help the cell survive. This level of control ensures that cells can quickly adapt their protein production based on external and internal signals.

3.7 POST-TRANSLATIONAL CONTROL: MODIFYING PROTEINS AFTER SYNTHESIS

Post-translational control refers to the regulation of protein activity and stability after translation has occurred. Many proteins undergo post-translational modifications, such as phosphorylation, methylation, acetylation, and ubiquitination, which can alter their activity, localization, stability, or interactions with other molecules. For instance, phosphorylation by kinases can activate or deactivate enzymes and signal transduction proteins, providing a rapid and reversible means of controlling protein function. Ubiquitination, where proteins are tagged with ubiquitin molecules, marks proteins for degradation by the proteasome. This system enables cells to regulate protein levels and remove damaged or misfolded proteins, preventing cellular stress and maintaining homeostasis. Post-translational control is thus crucial for dynamically adjusting protein functions and ensuring cellular responses to various stimuli.

3.8 SIGNAL-DEPENDENT GENE REGULATION

Gene expression can also be regulated in response to external signals. Signal transduction pathways, which often involve a series of protein phosphorylation events, relay information from the cell surface to the nucleus, ultimately affecting gene transcription. For example, in response to stress or hormones, specific signaling pathways can activate transcription factors that modify gene expression to produce proteins needed for adaptation.

One well-studied example is the response to hormones like estrogen and cortisol, which bind to nuclear receptors and activate transcription of genes involved in growth, metabolism, and stress response. Similarly, in immune responses, inflammatory signals activate transcription factors that regulate genes involved in pathogen defense. This ability to regulate gene expression based on environmental signals is vital for survival and adaptation in changing conditions.

3.9 THE IMPORTANCE OF GENE EXPRESSION CONTROL IN DEVELOPMENT AND DISEASE

The precise regulation of gene expression is essential for proper development and differentiation. During development, specific genes are activated or repressed to guide cells into becoming specialized types, such as muscle, nerve, or skin cells. Any disruption in this process can lead to developmental abnormalities and diseases.

Gene expression control also plays a critical role in cancer, where mutations in regulatory proteins like transcription factors, chromatin remodelers, or signaling molecules can lead to uncontrolled cell growth. Abnormal gene regulation is a common feature in many diseases, including metabolic disorders, immune deficiencies, and neurological conditions. Understanding the mechanisms of gene regulation not only sheds light on fundamental biology but also provides potential therapeutic targets for treating diseases.

3.10 CONCLUSION

The control of gene expression is a highly complex and finely tuned process involving multiple levels of regulation, from DNA to mRNA to protein. This intricate regulation allows organisms to respond dynamically to environmental changes, carry out specialized functions, and maintain cellular health. Each level of control—from transcriptional regulation to post-translational modifications—plays a unique role in ensuring that genes are expressed at the right time, in the right place, and at the right intensity. Advances in molecular biology have deepened our understanding of gene regulation, enabling us to explore new treatments for diseases linked to gene expression abnormalities. As research progresses, the study of gene expression control will continue to illuminate the mysteries of cellular function and provide new insights into human health and disease.

CHAPTER 4
BASIC TOOLS OF RECOMBINANT DNA

Recombinant DNA (rDNA) technology is a powerful tool in modern biotechnology, enabling scientists to manipulate genes and create DNA sequences that would not otherwise be found in nature. It involves combining DNA from different organisms, often to express specific genes in a host cell to produce desired proteins, study gene function, or even develop treatments for genetic diseases. This technique has revolutionized fields like medicine, agriculture, and environmental science. The success of recombinant DNA technology depends on a variety of molecular tools that allow precise manipulation of genetic material.

4.1 RESTRICTION ENZYMES: THE MOLECULAR SCISSORS

Restriction enzymes, also known as restriction endonucleases, are proteins that cut DNA at specific sequences, playing an essential role in molecular biology and recombinant DNA technology. These enzymes, naturally found in bacteria, act as a defense mechanism against viral DNA, allowing bacteria to "restrict" the growth of viruses by cleaving foreign DNA. Over the decades, scientists have harnessed these enzymes to manipulate and study DNA, making restriction enzymes foundational tools for genetic engineering, cloning, and DNA analysis. Restriction enzymes are a cornerstone of molecular biology and genetics. Their ability to cut DNA at precise sites has enabled countless breakthroughs in genetic research, biotechnology, and medicine. Though newer tools like CRISPR have expanded the possibilities of genome editing, restriction enzymes remain indispensable for basic DNA manipulation tasks, demonstrating the power and simplicity of nature's own molecular tools.

4.1.1 Discovery and Importance of Restriction Enzymes

The discovery of restriction enzymes in the 1960s revolutionized genetic research. Scientists realized that certain bacterial enzymes could cut DNA in a predictable manner, opening new avenues for DNA manipulation. Werner Arber, Hamilton Smith, and Daniel Nathans were awarded the Nobel Prize in 1978 for their pioneering work in identifying and characterizing restriction enzymes. These enzymes enabled researchers to cut DNA from any organism into precise fragments, enabling the analysis of genetic material, sequencing, and the construction of recombinant DNA.

4.1.2 Types of Restriction Enzymes

Restriction enzymes are classified into several types based on their structure, recognition sites, and cleavage patterns. The main types are:

- **Type I**: These enzymes cut DNA at random locations far from their recognition sites. They require ATP and other cofactors, making them less predictable for laboratory applications.
- **Type II**: These are the most widely used restriction enzymes in molecular biology. They recognize specific DNA sequences (usually 4-8 base pairs in length) and cleave within or very close to these sequences. Type II enzymes are straightforward and predictable, making them invaluable for cloning and DNA manipulation.
- **Type III**: These enzymes cut DNA a short distance away from their recognition sites and also require ATP, but they are less commonly used in research.

4.1.3 Recognition Sequences and Cleavage Patterns

Each restriction enzyme has a unique recognition sequence, which is typically palindromic (reads the same forward and backward). For example, **EcoRI**, a commonly used enzyme, recognizes the sequence **GAATTC** and cuts between G and A on each strand. Depending on the enzyme, the cut may result in "sticky" or "blunt" ends:

- **Sticky Ends**: Some enzymes, like EcoRI, create overhangs or "sticky ends" when they cut DNA. Sticky ends have unpaired nucleotides that can easily form hydrogen bonds with complementary sequences, making them useful for joining DNA fragments from different sources.
- **Blunt Ends**: Other enzymes, like **SmaI**, cut DNA straight across, resulting in "blunt ends" with no overhangs. Although less versatile than sticky ends for cloning, blunt ends can still be used to join DNA fragments with the help of additional enzymes.

4.1.4 Applications of Restriction Enzymes

Restriction enzymes have a variety of applications in molecular biology:

- **Genetic Cloning**: Restriction enzymes are essential for cloning, as they allow scientists to cut DNA from one organism and insert it into plasmids or vectors. This creates recombinant DNA, which can then be introduced into host cells to study gene function or produce proteins.
- **Genetic Mapping and Sequencing**: By cutting DNA into fragments, restriction enzymes enable the mapping of genes on chromosomes.

The fragments produced can be sequenced and reassembled, aiding in genome analysis and gene identification.

- **DNA Fingerprinting:** In forensic science, restriction enzymes are used to analyze genetic material by creating a unique DNA "fingerprint" for individuals. This is useful in paternity testing and criminal investigations.
- **Gene Therapy:** Restriction enzymes are key in gene therapy, where faulty genes are replaced with functional copies. They allow scientists to insert therapeutic genes into vectors that can deliver the new genes to a patient's cells.

4.1.5 Limitations of Restriction Enzymes

Despite their widespread use, restriction enzymes have some limitations:

- **Specificity of Recognition Sites:** Restriction enzymes are limited by the need for specific recognition sequences, which may not always be present in the desired locations. This can make it challenging to work with certain DNA sequences.
- **Star Activity:** Under non-optimal conditions (such as high glycerol concentration, low ionic strength, or high pH), some enzymes exhibit "star activity," where they cleave at sites similar to, but not exactly matching, their recognition sequence. This can lead to unintended DNA modifications.
- **Inability to Edit at the Nucleotide Level:** Unlike newer gene-editing technologies like CRISPR-Cas9, restriction enzymes cannot alter individual nucleotides; they only cut at specific sequences. This limits their utility in precise genome editing applications.

4.1.6 Future and Alternative DNA Manipulation Tools

The field of genetic research has evolved with the development of other tools like **CRISPR-Cas9**, which allows for precise, programmable DNA editing. While CRISPR provides a more targeted approach, restriction enzymes remain invaluable for routine DNA manipulation and cloning. They continue to be an essential tool in labs worldwide, laying the groundwork for more advanced techniques.

4.2 DNA LIGASE: THE MOLECULAR GLUE

DNA ligase, often referred to as the "molecular glue," is a crucial enzyme in molecular biology responsible for joining DNA strands together by forming bonds between the sugar-phosphate backbones of DNA molecules. This enzyme plays an essential role in both natural DNA replication and repair, as well as in recombinant

DNA technology, where scientists use it to construct new DNA molecules by linking different fragments. By sealing breaks in DNA, DNA ligase maintains the stability and integrity of genetic information, ensuring the continuity of life at the molecular level. DNA ligase, as the "molecular glue" of genetic research, is indispensable in both natural DNA maintenance and laboratory applications. Its ability to form bonds between DNA strands has laid the foundation for molecular biology techniques, from genetic engineering to cloning and gene therapy. The enzyme's versatility, combined with newer gene-editing technologies like CRISPR, continues to expand the possibilities in genomics, medicine, and biotechnology. DNA ligase thus remains one of the fundamental tools that allow scientists to explore and manipulate the genetic code.

4.2.1 Role of DNA Ligase in Cells

In cells, DNA ligase is fundamental for DNA replication and repair. During DNA replication, the enzyme helps in sealing the "nicks" or breaks in the sugar-phosphate backbone of the newly synthesized lagging strand, which is created as short segments called Okazaki fragments. Each Okazaki fragment requires the action of DNA ligase to form a continuous strand, essential for producing an intact daughter DNA molecule. DNA ligase is also critical for DNA repair mechanisms. It is involved in fixing breaks that arise from exposure to radiation, chemicals, or other environmental factors. By joining DNA fragments, DNA ligase restores the structural integrity of DNA molecules, thus preventing mutations that could lead to various diseases, including cancer.

4.2.2 Types of DNA Ligase

There are several types of DNA ligases found in nature, each serving specific roles:

- **E. coli DNA Ligase:** This type is commonly found in bacteria and is often used in laboratory settings. It relies on NAD+ as a cofactor for its action.
- **T4 DNA Ligase:** Derived from bacteriophage T4, this type of DNA ligase is the most widely used in molecular biology. T4 DNA ligase can join both sticky and blunt ends of DNA and uses ATP as a cofactor, making it versatile for cloning and DNA manipulation.
- **Human DNA Ligase:** In human cells, DNA ligase has multiple forms, with DNA ligase I involved in DNA replication, while ligases III and IV play roles in DNA repair. Each ligase type has specialized functions suited to different cellular needs.

4.2.3 DNA Ligase Mechanism

The mechanism by which DNA ligase works involves three main steps:

- **Adenylation:** DNA ligase becomes activated when it attaches an AMP molecule from ATP (or NAD+ in bacterial ligases) to itself. This step primes the enzyme to transfer AMP to the DNA molecule it will act upon.

- **AMP Transfer to DNA:** The enzyme transfers the AMP molecule to the 5' end of a DNA strand with a nick. This prepares the DNA for the final bonding step.
- **Bond Formation and AMP Release:** The enzyme catalyzes the formation of a phosphodiester bond between the 3' hydroxyl group of one nucleotide and the 5' phosphate group of another, sealing the nick in the DNA strand. The AMP is then released, and the DNA ligase is ready to catalyze another reaction.

4.2.4 Applications in Recombinant DNA Technology

DNA ligase is invaluable in genetic engineering and recombinant DNA technology. Scientists use DNA ligase to join DNA fragments, enabling them to insert genes or other DNA segments into plasmids or vectors. These recombinant molecules can then be introduced into host cells for cloning, protein production, or further genetic study. Applications include:

- **Cloning:** In genetic cloning, DNA ligase is used to insert specific genes into plasmids. This allows researchers to amplify or express desired genes, such as those for insulin production or vaccine development.
- **Genetic Engineering:** DNA ligase is key in creating genetically modified organisms (GMOs) by inserting foreign DNA into the genomes of plants, animals, or bacteria. These modifications lead to traits like pest resistance in crops or the production of useful proteins in bacteria.
- **Gene Therapy:** DNA ligase has applications in gene therapy, where therapeutic genes are inserted into patients' cells to replace faulty genes. By joining the therapeutic gene with a vector, scientists can deliver the gene to target cells, potentially treating genetic disorders.

4.2.5 Advantages and Limitations of DNA Ligase

Advantages

- **Versatile in Laboratory Applications:** T4 DNA ligase, in particular, can join both sticky and blunt ends, making it a flexible tool for different DNA manipulation tasks.
- **Precision:** DNA ligase allows for precise recombination of DNA segments, essential for accurate cloning and genetic engineering.

Limitations

- **Requires ATP or NAD+ for Activation:** DNA ligase activity depends on specific cofactors (ATP for T4 ligase, NAD+ for bacterial ligase), making reaction conditions critical for its function.

- **Inefficient Blunt-End Ligation:** Ligation of blunt ends is less efficient than sticky ends, sometimes requiring higher concentrations of ligase or optimized reaction conditions.

4.2.6 DNA Ligase and CRISPR

While CRISPR-Cas9 technology has advanced gene editing by enabling precise DNA cutting, DNA ligase remains essential in the DNA repair process. After CRISPR introduces double-stranded breaks, DNA ligase can help facilitate the repair and insertion of new genetic material. This collaboration between CRISPR and DNA ligase has led to more sophisticated genome editing approaches, enabling the development of targeted therapies and more precise genetic modifications.

4.3 PLASMIDS: VECTORS FOR DNA TRANSFER

A plasmid is a small, extrachromosomal DNA molecule within a cell that is physically separated from chromosomal DNA and can replicate independently. They are most commonly found as small circular, double-stranded DNA molecules in bacteria; however, plasmids are sometimes present in archaea and eukaryotic organisms. Plasmids often carry useful genes, such as antibiotic resistance and virulence. While chromosomes are large and contain all the essential genetic information for living under normal conditions, plasmids are usually very small and contain additional genes for special circumstances.

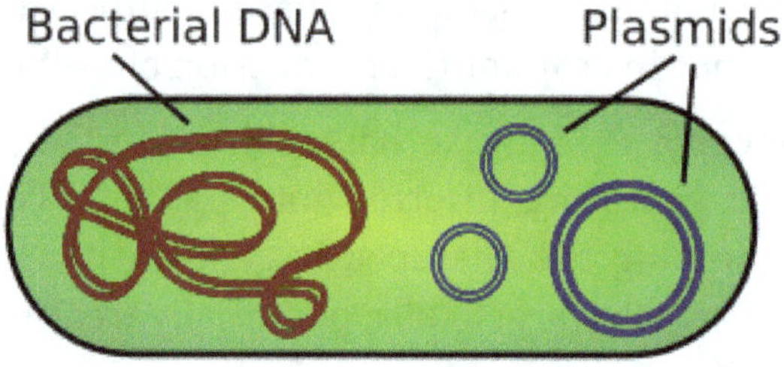

Fig: Illustration of a Bacterium Showing Chromosomal DNA and Plasmids

Artificial plasmids are widely used as vectors in molecular cloning, serving to drive the replication of recombinant DNA sequences within host organisms. In the laboratory, plasmids may be introduced into a cell via transformation. Synthetic plasmids are available for procurement over the internet.

Plasmids are considered *replicons*, units of DNA capable of replicating autonomously within a suitable host. However, plasmids, like viruses, are not generally classified as life. Plasmids are transmitted from one bacterium to another (even of another species) mostly through conjugation. This host-to-host transfer of genetic material is one mechanism of horizontal gene transfer, and plasmids are considered part of the mobilome. Unlike viruses, which encase their genetic material in a protective protein coat called a capsid, plasmids are "naked" DNA and do not encode genes necessary to encase the genetic material for transfer to a new host; however, some classes of plasmids encode the conjugative "sex" pilus necessary for their own transfer. Plasmids vary in size from 1 to over 400 kbp, and the number of identical plasmids in a single cell can range from one up to thousands.

4.3.1 History

The term *plasmid* was coined in 1952 by the American molecular biologist Joshua Lederberg to refer to "any extrachromosomal hereditary determinant." The term's early usage included any bacterial genetic material that exists extrachromosomally for at least part of its replication cycle, but because that description includes bacterial viruses, the notion of plasmid was refined over time to refer to genetic elements that reproduce autonomously. Later in 1968, it was decided that the term plasmid should be adopted as the term for extrachromosomal genetic element, and to distinguish it from viruses, the definition was narrowed to genetic elements that exist exclusively or predominantly outside of the chromosome and can replicate autonomously.

4.3.2 Properties and Characteristics

In order for plasmids to replicate independently within a cell, they must possess a stretch of DNA that can act as an origin of replication. The self-replicating unit, in this case, the plasmid, is called a replicon. A typical bacterial replicon may consist of a number of elements, such as the gene for plasmid-specific replication initiation protein (Rep), repeating units called iterons, DnaA boxes, and an adjacent AT-rich region. Smaller plasmids make use of the host replicative enzymes to make copies of themselves, while larger plasmids may carry genes specific for the replication of those plasmids. A few types of plasmids can also insert into the host chromosome, and these integrative plasmids are sometimes referred to as episomes in prokaryotes.

Plasmids almost always carry at least one gene. Many of the genes carried by a plasmid are beneficial for the host cells, for example: enabling the host cell to survive in an environment that would otherwise be lethal or restrictive for growth. Some of these genes encode traits for antibiotic resistance or resistance to heavy metal, while others may produce virulence factors that enable a bacterium to colonize a host and overcome its defences or have specific metabolic functions that allow the bacterium to utilize a particular nutrient, including the ability to degrade recalcitrant or toxic organic compounds. Plasmids can also provide bacteria with the ability to fix nitrogen. Some plasmids, called cryptic plasmids, play a crucial role in horizontal genes transfer, since they carry antibiotic-resistance genes. Thus they are important factors in spreading resistance, which can result in antibiotic treatment failures.

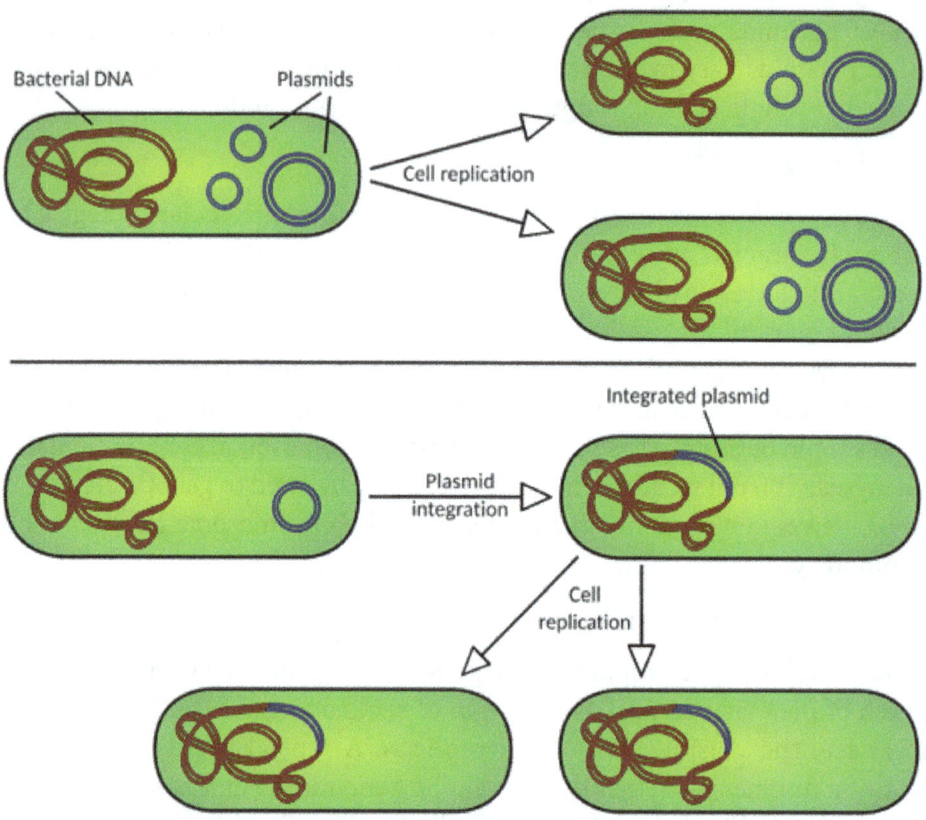

Fig: There are Two Types of Plasmid Integration into a Host Bacteria: Non-Integrating Plasmids Replicate as with the Top Instance, whereas Episomes, the Lower Example, can Integrate into the Host Chromosome

Naturally occurring plasmids vary greatly in their physical properties. Their size can range from very small mini-plasmids of less than 1-kilobase pairs (kbp) to very large megaplasmids of several megabase pairs (Mbp). At the upper end, little differs between a megaplasmid and a minichromosome. Plasmids are generally circular, but examples of linear plasmids are also known. These linear plasmids require specialized mechanisms to replicate their ends.

Plasmids may be present in an individual cell in varying number, ranging from one to several hundreds. The normal number of copies of plasmid that may be found in a single cell is called the plasmid copy number, and is determined by how the replication initiation is regulated and the size of the molecule. Larger plasmids tend to have lower copy numbers. Low-copy-number plasmids that exist only as one or a few copies in each bacterium are, upon cell division, in danger of being lost in one of the segregating bacteria. Such single-copy plasmids have systems that attempt to actively distribute a copy to both daughter cells. These systems, which include the parABS system and parMRC system, are often referred to as the partition system or partition function of a plasmid.

4.3.3 Classifications and Types

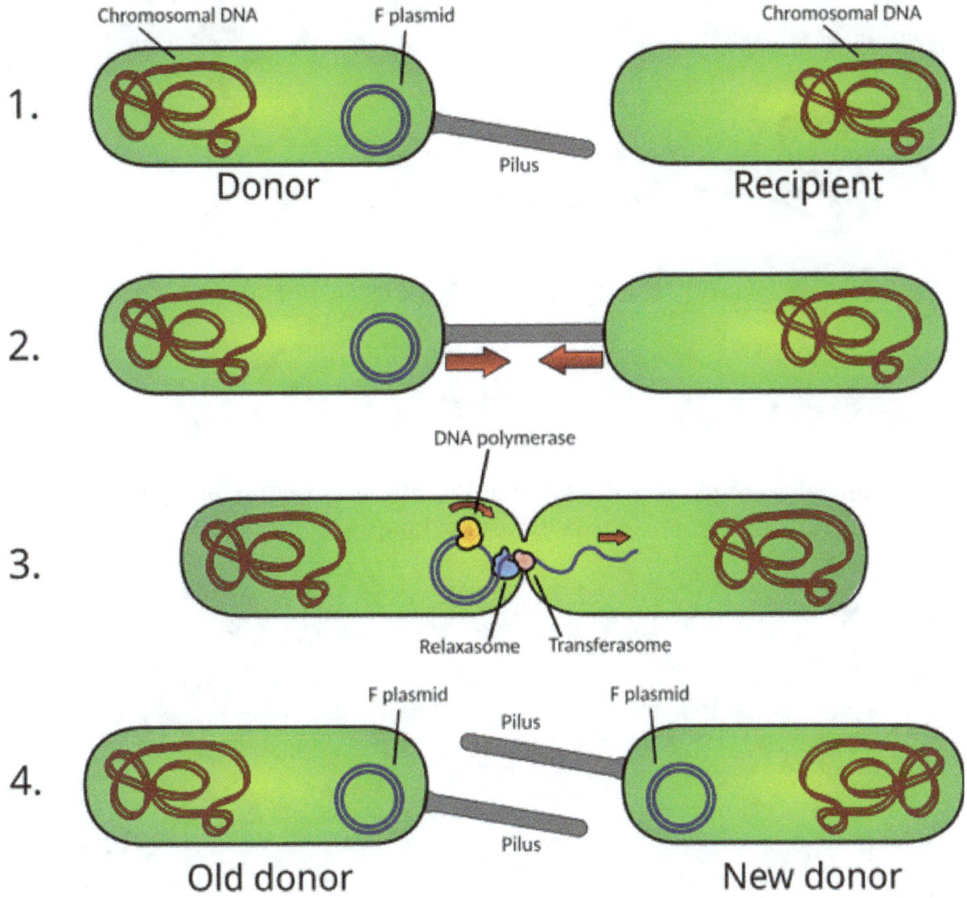

Fig: Overview of Bacterial Conjugation

Plasmids may be classified in a number of ways. Plasmids can be broadly classified into conjugative plasmids and non-conjugative plasmids. Conjugative plasmids contain a set of transfer genes which promote sexual conjugation between different cells. In the complex process of conjugation, plasmids may be transferred from one bacterium to another via sex pili encoded by some of the transfer genes. Non-conjugative plasmids are incapable of initiating conjugation, hence they can be transferred only with the assistance of conjugative plasmids. An intermediate class of plasmids are mobilizable, and carry only a subset of the genes required for transfer. They can parasitize a conjugative plasmid, transferring at high frequency only in its presence.

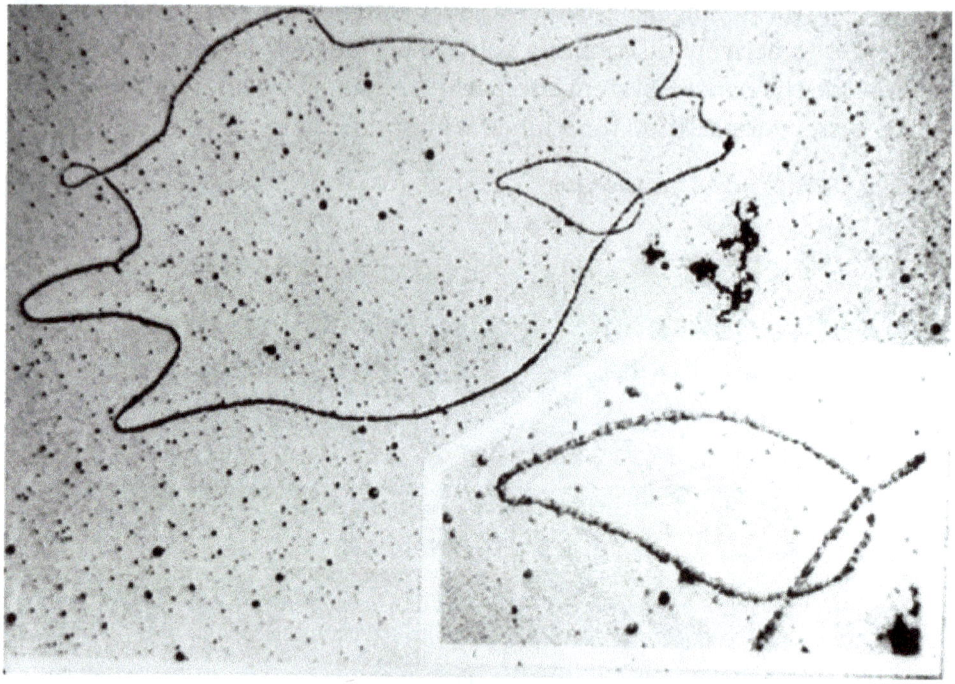

Fig: Electron Micrograph of a DNA Fiber Bundle, Presumably of a Single Bacterial Chromosome Loop

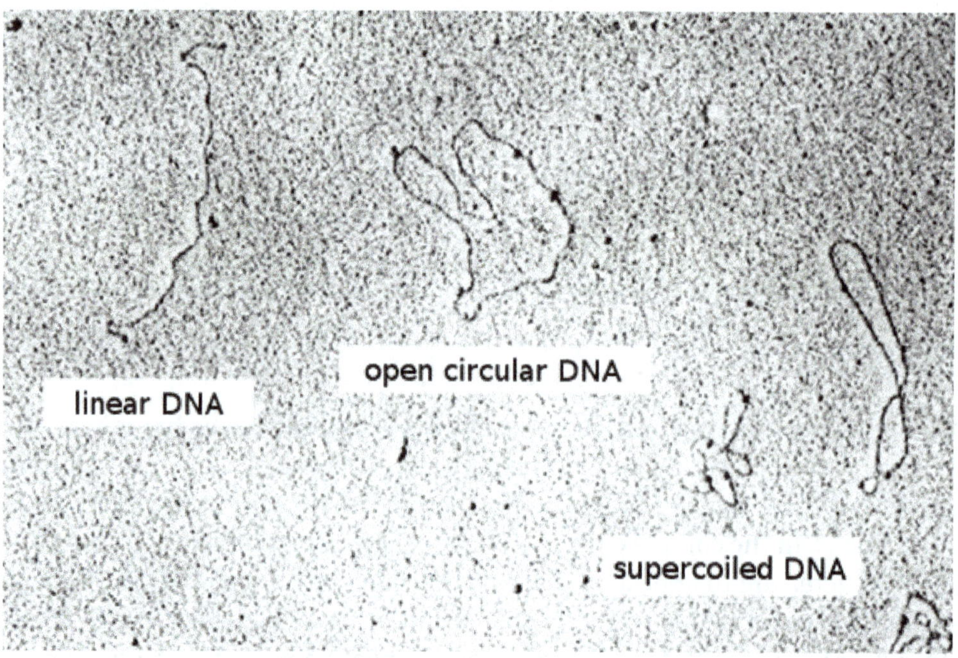

Fig: Electron Micrograph of a Bacterial DNA Plasmid (Chromosome Fragment)

Plasmids can also be classified into incompatibility groups. A microbe can harbour different types of plasmids, but different plasmids can only exist in a single bacterial cell if they are compatible. If two plasmids are not compatible, one or the other will be rapidly lost from the cell. Different plasmids may therefore be assigned to different incompatibility groups depending on whether they can coexist together. Incompatible plasmids (belonging to the same incompatibility group) normally share the same replication or partition mechanisms and can thus not be kept together in a single cell.

Another way to classify plasmids is by function. There are five main classes:

- **Fertility F-plasmids**, which contain *tra* genes. They are capable of conjugation and result in the expression of sex pili.
- **Resistance (R) plasmids**, which contain genes that provide resistance against antibiotics or antibacterial agents. Historically known as R-factors, before the nature of plasmids was understood.
- **Col plasmids**, which contain genes that code for bacteriocins, proteins that can kill other bacteria.
- **Degradative plasmids**, which enable the digestion of unusual substances, e.g., toluene and salicylic acid.
- **Virulence plasmids**, which turn the bacterium into a pathogen. e.g. Ti plasmid in *Agrobacterium tumefaciens*

RNA Plasmids

Although most plasmids are double-stranded DNA molecules, some consist of single-stranded DNA, or predominantly double-stranded RNA. RNA plasmids are non-infectious extrachromosomal linear RNA replicons, both encapsidated and unencapsidated, which have been found in fungi and various plants, from algae to land plants. In many cases, however, it may be difficult or impossible to clearly distinguish RNA plasmids from RNA viruses and other infectious RNAs.

Chromids

Chromids are elements that exist at the boundary between a chromosome and a plasmid, found in about 10% of bacterial species sequenced by 2009. These elements carry core genes and have codon usage similar to the chromosome, yet use a plasmid-type replication mechanism such as the low copy number RepABC. As a result, they have been variously classified as minichromosomes or megaplasmids in the past. In *Vibrio*, the bacterium synchronizes the replication of the chromosome and chromid by a conserved genome size ratio.

4.3.4 Vectors

Artificially constructed plasmids may be used as vectors in genetic engineering. These plasmids serve as important tools in genetics and biotechnology labs, where they are commonly used to clone and amplify (make many copies of) or express particular genes. A wide variety of plasmids are commercially available for such uses. The gene to be replicated is normally inserted into a plasmid that typically contains a number of features for their use. These include a gene that confers resistance to particular antibiotics (ampicillin is most frequently used for bacterial strains), an origin of replication to allow the bacterial cells to replicate the plasmid DNA, and a suitable site for cloning (referred to as a multiple cloning site).

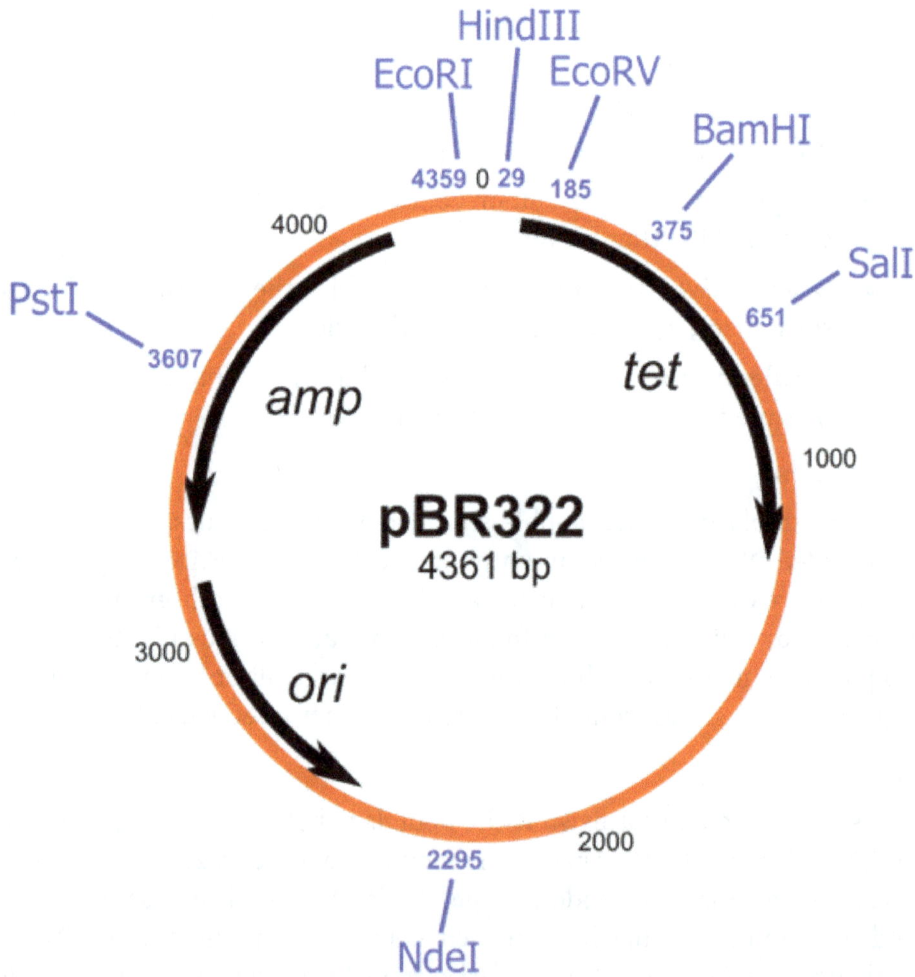

Fig: A Schematic Representation of the pBR322 Plasmid, One of the First Plasmids to be Used Widely as a Cloning Cector.

DNA structural instability can be defined as a series of spontaneous events that culminate in an unforeseen rearrangement, loss, or gain of genetic material.

Such events are frequently triggered by the transposition of mobile elements or by the presence of unstable elements such as non-canonical (non-B) structures. Accessory regions pertaining to the bacterial backbone may engage in a wide range of structural instability phenomena. Well-known catalysts of genetic instability include direct, inverted, and tandem repeats, which are known to be conspicuous in a large number of commercially available cloning and expression vectors. Insertion sequences can also severely impact plasmid function and yield, by leading to deletions and rearrangements, activation, down-regulation or inactivation of neighboring gene expression. Therefore, the reduction or complete elimination of extraneous noncoding backbone sequences would pointedly reduce the propensity for such events to take place, and consequently, the overall recombinogenic potential of the plasmid.

Cloning

Plasmids are the most-commonly used bacterial cloning vectors. These cloning vectors contain a site that allows DNA fragments to be inserted, for example a multiple cloning site or polylinker which has several commonly used restriction sites to which DNA fragments may be ligated. After the gene of interest is inserted, the plasmids are introduced into bacteria by a process called transformation. These plasmids contain a selectable marker, usually an antibiotic resistance gene, which confers on the bacteria an ability to survive and proliferate in a selective growth medium containing the particular antibiotics. The cells after transformation are exposed to the selective media, and only cells containing the plasmid may survive. In this way, the antibiotics act as a filter to select only the bacteria containing the plasmid DNA. The vector may also contain other marker genes or reporter genes to facilitate selection of plasmids with cloned inserts. Bacteria containing the plasmid can then be grown in large amounts, harvested, and the plasmid of interest may then be isolated using various methods of plasmid preparation.

A plasmid cloning vector is typically used to clone DNA fragments of up to 15 kbp. To clone longer lengths of DNA, lambda phage with lysogeny genes deleted, cosmids, bacterial artificial chromosomes, or yeast artificial chromosomes are used.

Protein Production

Another major use of plasmids is to make large amounts of proteins. In this case, researchers grow bacteria containing a plasmid harboring the gene of interest. Just as the bacterium produces proteins to confer its antibiotic resistance, it can also be induced to produce large amounts of proteins from the inserted gene. This is a cheap and easy way of mass-producing the protein the gene codes for, for example, insulin.

Gene Therapy

Plasmids may also be used for gene transfer as a potential treatment in gene therapy so that it may express the protein that is lacking in the cells. Some forms of gene therapy require the insertion of therapeutic genes at pre-selected chromosomal target sites within the human genome. Plasmid vectors are one of many approaches that could be used for this purpose. Zinc finger nucleases (ZFNs) offer a way to cause a site-specific double-strand break to the DNA genome and cause homologous recombination. Plasmids encoding ZFN could help deliver a therapeutic gene to a specific site so that cell damage, cancer-causing mutations, or an immune response is avoided.

Disease Models

Plasmids were historically used to genetically engineer the embryonic stem cells of rats to create rat genetic disease models. The limited efficiency of plasmid-based techniques precluded their use in the creation of more accurate human cell models. However, developments in adeno-associated virus recombination techniques, and zinc finger nucleases, have enabled the creation of a new generation of isogenic human disease models.

Biogenetic Gene Cluster (BGC)

Plasmids assist in transporting biogenetic gene clusters - a set of gene that contain all the necessary enzymes that lead to the production of special metabolites (formally known as secondary metabolite). A benefit of using plasmids to transfer BGC is demonstrated by using a suitable host that can mass produce specialized metabolites, some of these molecules are able to control microbial population. Plasmids can contain and express several BGCs with a few plasmids known to be exclusive for transferring BGCs. BGC's can also be transfers to the host organism's chromosome, utilizing a plasmid vector, which allows for studies in gene knockout experiments. By using plasmids for the uptake of BGCs, microorganisms can gain an advantage as production is not limited to antibiotic resistant biosynthesis genes but the production of toxins/antitoxins.

4.3.5 Episomes

The term *episome* was introduced by François Jacob and Élie Wollman in 1958 to refer to extra-chromosomal genetic material that may replicate autonomously or become integrated into the chromosome. Since the term was introduced, however, its use has changed, as *plasmid* has become the preferred term for autonomously replicating extrachromosomal DNA. At a 1968 symposium in London some participants suggested that the term *episome* be abandoned, although others continued to use the term with a shift in meaning.

Today, some authors use *episome* in the context of prokaryotes to refer to a plasmid that is capable of integrating into the chromosome. The integrative plasmids may be replicated and stably maintained in a cell through multiple generations, but at some stage, they will exist as an independent plasmid molecule. In the context of eukaryotes, the term *episome* is used to mean a non-integrated extrachromosomal closed circular DNA molecule that may be replicated in the nucleus. Viruses are the most common examples of this, such as herpesviruses, adenoviruses, and polyomaviruses, but some are plasmids. Other examples include aberrant chromosomal fragments, such as double minute chromosomes, that can arise during artificial gene amplifications or in pathologic processes (e.g., cancer cell transformation). Episomes in eukaryotes behave similarly to plasmids in prokaryotes in that the DNA is stably maintained and replicated with the host cell. Cytoplasmic viral episomes (as in poxvirus infections) can also occur. Some episomes, such as herpesviruses, replicate in a rolling circle mechanism, similar to bacteriophages (bacterial phage viruses). Others replicate through a bidirectional replication mechanism (*Theta type* plasmids). In either case, episomes remain physically separate from host cell chromosomes. Several cancer viruses, including Epstein-Barr virus and Kaposi's sarcoma-associated herpesvirus, are maintained as latent, chromosomally distinct episomes in cancer cells, where the viruses express oncogenes that promote cancer cell proliferation. In cancers, these episomes passively replicate together with host chromosomes when the cell divides. When these viral episomes initiate lytic replication to generate multiple virus particles, they generally activate cellular innate immunity defense mechanisms that kill the host cell.

4.3.6 Plasmid Maintenance

Some plasmids or microbial hosts include an addiction system or postsegregational killing system (PSK), such as the hok/sok (host killing/suppressor of killing) system of plasmid R1 in *Escherichia coli*. This variant produces both a long-lived poison and a short-lived antidote. Several types of plasmid addiction systems (toxin/antitoxin, metabolism-based, ORT systems) were described in the literature and used in biotechnical (fermentation) or biomedical (vaccine therapy) applications. Daughter cells that retain a copy of the plasmid survive, while a daughter cell that fails to inherit the plasmid dies or suffers a reduced growth-rate because of the lingering poison from the parent cell. Finally, the overall productivity could be enhanced.

In contrast, plasmids used in biotechnology, such as pUC18, pBR322 and derived vectors, hardly ever contain toxin-antitoxin addiction systems, and therefore need to be kept under antibiotic pressure to avoid plasmid loss.

4.3.7 Plasmids in Nature

Yeast Plasmids

Yeasts naturally harbour various plasmids. Notable among them are 2 µm plasmids—small circular plasmids often used for genetic engineering of yeast—and linear pGKL plasmids from *Kluyveromyces lactis*, that are responsible for killer phenotypes.

Other types of plasmids are often related to yeast cloning vectors that include:

- **Yeast integrative plasmid (YIp),** yeast vectors that rely on integration into the host chromosome for survival and replication, and are usually used when studying the functionality of a solo gene or when the gene is toxic. Also connected with the gene URA3, that codes an enzyme related to the biosynthesis of pyrimidine nucleotides (T, C);
- **Yeast Replicative Plasmid (YRp)**, which transport a sequence of chromosomal DNA that includes an origin of replication. These plasmids are less stable, as they can be lost during budding.

Plant Mitochondrial Plasmids

The mitochondria of many higher plants contain self-replicating, extra-chromosomal linear or circular DNA molecules which have been considered to be plasmids. These can range from 0.7 kb to 20 kb in size. The plasmids have been generally classified into two categories- circular and linear. Circular plasmids have been isolated and found in many different plants, with those in *Vicia faba and Chenopodium album* being the most studied and whose mechanism of replication is known. The circular plasmids can replicate using the θ model of replication (as in *Vicia faba*) and through rolling circle replication (as in *C.album*). Linear plasmids have been identified in some plant species such as *Beta vulgaris, Brassica napus, Zea mays*, etc. but are rarer than their circular counterparts.

The function and origin of these plasmids remains largely unknown. It has been suggested that the circular plasmids share a common ancestor, some genes in the mitochondrial plasmid have counterparts in the nuclear DNA suggesting inter-compartment exchange. Meanwhile, the linear plasmids share structural similarities such as invertrons with viral DNA and fungal plasmids, like fungal plasmids they also have low GC content, these observations have led to some hypothesizing that these linear plasmids have viral origins, or have ended up in plant mitochondria through horizontal gene transfer from pathogenic fungi.

4.3.8 Study of Plasmids

Plasmid DNA Extraction

Plasmids are often used to purify a specific sequence, since they can easily be purified away from the rest of the genome. For their use as vectors, and for molecular cloning, plasmids often need to be isolated. There are several methods to isolate plasmid DNA from bacteria, ranging from the plasmid extraction kits (miniprep to the maxiprep or bulkprep), alkaline lysis, enzymatic lysis, and mechanical lysis . The former can be used to quickly find out whether the plasmid is correct in any of several bacterial clones. The yield is a small amount of impure plasmid DNA, which is sufficient for analysis by restriction digest and for some cloning techniques.

In the latter, much larger volumes of bacterial suspension are grown from which a maxi-prep can be performed. In essence, this is a scaled-up miniprep followed by additional purification. This results in relatively large amounts (several hundred micrograms) of very pure plasmid DNA. Many commercial kits have been created to perform plasmid extraction at various scales, purity, and levels of automation.

Conformations

Plasmid DNA may appear in one of five conformations, which (for a given size) run at different speeds in a gel during electrophoresis.

- **Nicked open-circular** DNA has one strand cut.
- **Relaxed circular** DNA is fully intact with both strands uncut but has been enzymatically *relaxed* (supercoils removed). This can be modeled by letting a twisted extension cord unwind and relax and then plugging it into itself.
- **Linear** DNA has free ends, either because both strands have been cut or because the DNA was linear *in vivo*. This can be modeled with an electrical extension cord that is not plugged into itself.
- **Supercoiled (or covalently closed-circular)** DNA is fully intact with both strands uncut, and with an integral twist, resulting in a compact form. This can be modeled by twisting an extension cord and then plugging it into itself.
- **Supercoiled denatured** DNA is similar to *supercoiled DNA*, but has unpaired regions that make it slightly less compact; this can result from excessive alkalinity during plasmid preparation.

The rate of migration for small linear fragments is directly proportional to the voltage applied at low voltages. At higher voltages, larger fragments migrate at continuously increasing yet different rates. Thus, the resolution of a gel decreases with increased voltage.

At a specified, low voltage, the migration rate of small linear DNA fragments is a function of their length. Large linear fragments (over 20 kb or so) migrate at a certain fixed rate regardless of length. This is because the molecules 'respirate', with the bulk of the molecule following the leading end through the gel matrix. Restriction digests are frequently used to analyse purified plasmids. These enzymes specifically break the DNA at certain short sequences. The resulting linear fragments form 'bands' after gel electrophoresis. It is possible to purify certain fragments by cutting the bands out of the gel and dissolving the gel to release the DNA fragments. Because of its tight conformation, supercoiled DNA migrates faster through a gel than linear or open-circular DNA.

Software for Bioinformatics and Design

The use of plasmids as a technique in molecular biology is supported by bioinformatics software. These programs record the DNA sequence of plasmid vectors, help to predict cut sites of restriction enzymes, and to plan manipulations. Examples of software packages that handle plasmid maps are ApE, Clone Manager, GeneConstructionKit, Geneious, Genome Compiler, LabGenius, Lasergene, MacVector, pDraw32, Serial Cloner, UGENE, VectorFriends, Vector NTI, and WebDSV. These pieces of software help conduct entire experiments in silico before doing wet experiments.

Plasmid Collections

Many plasmids have been created over the years and researchers have given out plasmids to plasmid databases such as the non-profit organisations Addgene and BCCM/GeneCorner. One can find and request plasmids from those databases for research. Researchers also often upload plasmid sequences to the NCBI database, from which sequences of specific plasmids can be retrieved.

4.4 POLYMERASE CHAIN REACTION (PCR): AMPLIFYING DNA

The polymerase chain reaction (PCR) is a technique that allows researchers to make millions of copies of a specific DNA sequence in a short time. PCR has revolutionized recombinant DNA technology by making it easier to obtain sufficient quantities of DNA for further analysis or manipulation. The process relies on a heat-stable DNA polymerase enzyme (such as Taq polymerase) and short DNA sequences called primers that target the region of interest. During PCR, the DNA sample is repeatedly heated and cooled through a series of cycles. The heating step denatures the DNA, separating the strands, while the cooling step allows primers to bind to complementary sequences on each strand. The DNA polymerase then extends the primers, synthesizing new DNA strands. Each cycle doubles the amount of DNA, quickly resulting in millions of copies of the target sequence. PCR is widely used in gene cloning, DNA sequencing, and diagnostic applications.

4.5 GEL ELECTROPHORESIS: VISUALIZING DNA FRAGMENTS

Gel electrophoresis is a technique used to separate DNA fragments based on size and charge. In recombinant DNA experiments, gel electrophoresis allows researchers to analyze and confirm the presence and size of DNA fragments obtained through restriction enzyme digestion or PCR. DNA samples are loaded into wells in a gel matrix, usually made of agarose, and an electric current is applied.

Since DNA is negatively charged, it migrates towards the positive electrode. Smaller DNA fragments move faster and travel farther through the gel, while larger fragments move more slowly. The result is a pattern of separated DNA bands that can be visualized using a DNA-binding dye, such as ethidium bromide. Gel electrophoresis is a critical tool for verifying that the correct DNA fragments have been obtained and for confirming successful cloning or PCR amplification.

4.6 DNA SEQUENCING: DETERMINING THE DNA SEQUENCE

DNA sequencing is the process of determining the nucleic acid sequence - the order of nucleotides in DNA. It includes any method or technology that is used to determine the order of the four bases: adenine, guanine, cytosine, and thymine. The advent of rapid DNA sequencing methods has greatly accelerated biological and medical research and discovery.

Knowledge of DNA sequences has become indispensable for basic biological research, DNA Genographic Projects and in numerous applied fields such as medical diagnosis, biotechnology, forensic biology, virology and biological systematics. Comparing healthy and mutated DNA sequences can diagnose different diseases including various cancers, characterize antibody repertoire, and can be used to guide patient treatment. Having a quick way to sequence DNA allows for faster and more individualized medical care to be administered, and for more organisms to be identified and cataloged.

The rapid speed of sequencing attained with modern DNA sequencing technology has been instrumental in the sequencing of complete DNA sequences, or genomes, of numerous types and species of life, including the human genome and other complete DNA sequences of many animal, plant, and microbial species.

The first DNA sequences were obtained in the early 1970s by academic researchers using laborious methods based on two-dimensional chromatography. Following the development of fluorescence-based sequencing methods with a DNA sequencer, DNA sequencing has become easier and orders of magnitude faster.

4.6.1 Applications

DNA sequencing may be used to determine the sequence of individual genes, larger genetic regions (i.e., clusters of genes or operons), full chromosomes, or entire genomes of any organism. DNA sequencing is also the most efficient way to indirectly sequence RNA or proteins (via their open reading frames). In fact, DNA sequencing has become a key technology in many areas of biology and other sciences such as medicine, forensics, and anthropology.

Molecular Biology

Sequencing is used in molecular biology to study genomes and the proteins they encode. Information obtained using sequencing allows researchers to identify changes in genes and noncoding DNA (including regulatory sequences), associations with diseases and phenotypes, and identify potential drug targets.

Evolutionary Biology

Since DNA is an informative macromolecule in terms of transmission from one generation to another, DNA sequencing is used in evolutionary biology to study how different organisms are related and how they evolved. In February 2021, scientists reported, for the first time, the sequencing of DNA from animal remains, a mammoth in this instance, over a million years old, the oldest DNA sequenced to date.

Metagenomics

The field of metagenomics involves identification of organisms present in a body of water, sewage, dirt, debris filtered from the air, or swab samples from organisms. Knowing which organisms are present in a particular environment is critical to research in ecology, epidemiology, microbiology, and other fields. Sequencing enables researchers to determine which types of microbes may be present in a microbiome, for example.

Virology

As most viruses are too small to be seen by a light microscope, sequencing is one of the main tools in virology to identify and study the virus. Viral genomes can be based in DNA or RNA. RNA viruses are more time-sensitive for genome sequencing, as they degrade faster in clinical samples. Traditional Sanger sequencing and next-generation sequencing are used to sequence viruses in basic and clinical research, as well as for the diagnosis of emerging viral infections, molecular epidemiology of viral pathogens, and drug-resistance testing. There are more than 2.3 million unique viral sequences in GenBank. Recently, NGS has surpassed traditional Sanger as the most popular approach for generating viral genomes. During the 1997 avian influenza outbreak, viral sequencing determined that the influenza sub-type originated through reassortment between quail and poultry. This led to legislation in Hong Kong that prohibited selling live quail and poultry together at market. Viral sequencing can also be used to estimate when a viral outbreak began by using a molecular clock technique.

Medicine

Medical technicians may sequence genes (or, theoretically, full genomes) from patients to determine if there is risk of genetic diseases. This is a form of genetic testing, though some genetic tests may not involve DNA sequencing. As of 2013 DNA sequencing was increasingly used to diagnose and treat rare diseases. As more and more genes are identified that cause rare genetic diseases, molecular diagnoses for patients become more mainstream. DNA sequencing allows clinicians to identify genetic diseases, improve disease management, provide reproductive counseling, and more effective therapies. Gene sequencing panels are used to identify multiple potential genetic causes of a suspected disorder.

Also, DNA sequencing may be useful for determining a specific bacteria, to allow for more precise antibiotics treatments, hereby reducing the risk of creating antimicrobial resistance in bacteria populations.

Forensic Investigation

DNA sequencing may be used along with DNA profiling methods for forensic identification and paternity testing. DNA testing has evolved tremendously in the last few decades to ultimately link a DNA print to what is under investigation. The DNA patterns in fingerprint, saliva, hair follicles, etc. uniquely separate each living organism from another. Testing DNA is a technique which can detect specific genomes in a DNA strand to produce a unique and individualized pattern.

DNA sequencing may be used along with DNA profiling methods for forensic identification and paternity testing, as it has evolved significantly over the past few decades to ultimately link a DNA print to what is under investigation. The DNA patterns in fingerprint, saliva, hair follicles, and other bodily fluids uniquely separate each living organism from another, making it an invaluable tool in the field of forensic science. The process of DNA testing involves detecting specific genomes in a DNA strand to produce a unique and individualized pattern, which can be used to identify individuals or determine their relationships.

The advancements in DNA sequencing technology have made it possible to analyze and compare large amounts of genetic data quickly and accurately, allowing investigators to gather evidence and solve crimes more efficiently. This technology has been used in various applications, including forensic identification, paternity testing, and human identification in cases where traditional identification methods are unavailable or unreliable. The use of DNA sequencing has also led to the development of new forensic techniques, such as DNA phenotyping, which allows investigators to predict an individual's physical characteristics based on their genetic data.

In addition to its applications in forensic science, DNA sequencing has also been used in medical research and diagnosis. It has enabled scientists to identify genetic mutations and variations that are associated with certain diseases and disorders, allowing for more accurate diagnoses and targeted treatments. Moreover, DNA sequencing has also been used in conservation biology to study the genetic diversity of endangered species and develop strategies for their conservation.

Furthermore, the use of DNA sequencing has also raised important ethical and legal considerations. For example, there are concerns about the privacy and security of genetic data, as well as the potential for misuse or discrimination based on genetic information. As a result, there are ongoing debates about the need for regulations and guidelines to ensure the responsible use of DNA sequencing technology.

Overall, the development of DNA sequencing technology has revolutionized the field of forensic science and has far-reaching implications for our understanding of genetics, medicine, and conservation biology.

4.6.2 The Four Canonical Bases

The canonical structure of DNA has four bases: thymine (T), adenine (A), cytosine (C), and guanine (G). DNA sequencing is the determination of the physical order of these bases in a molecule of DNA. However, there are many other bases that may be present in a molecule. In some viruses (specifically, bacteriophage), cytosine may be replaced by hydroxy methyl or hydroxy methyl glucose cytosine. In mammalian DNA, variant bases with methyl groups or phosphosulfate may be found. Depending on the sequencing technique, a particular modification, e.g., the 5mC (5 methyl cytosine) common in humans, may or may not be detected.

In almost all organisms, DNA is synthesized in vivo using only the 4 canonical bases; modification that occurs post replication creates other bases like 5 methyl C. However, some bacteriophage can incorporate a non standard base directly.

In addition to modifications, DNA is under constant assault by environmental agents such as UV and Oxygen radicals. At the present time, the presence of such damaged bases is not detected by most DNA sequencing methods, although PacBio has published on this.

4.6.3 History

Discovery of DNA Structure and Function

Deoxyribonucleic acid (DNA) was first discovered and isolated by Friedrich Miescher in 1869, but it remained under-studied for many decades because proteins, rather than DNA, were thought to hold the genetic blueprint to life. This situation changed after 1944 as a result of some experiments by Oswald Avery, Colin MacLeod, and Maclyn McCarty demonstrating that purified DNA could change one strain of bacteria into another. This was the first time that DNA was shown capable of transforming the properties of cells.

In 1953, James Watson and Francis Crick put forward their double-helix model of DNA, based on crystallized X-ray structures being studied by Rosalind Franklin. According to the model, DNA is composed of two strands of nucleotides coiled around each other, linked together by hydrogen bonds and running in opposite directions. Each strand is composed of four complementary nucleotides - adenine (A), cytosine (C), guanine (G) and thymine (T) - with an A on one strand always paired with T on the other, and C always paired with G. They proposed that such a structure allowed each strand to be used to reconstruct the other, an idea central to the passing on of hereditary information between generations.

The foundation for sequencing proteins was first laid by the work of Frederick Sanger who by 1955 had completed the sequence of all the amino acids in insulin, a small protein secreted by the pancreas. This provided the first conclusive evidence that proteins were chemical entities with a specific molecular pattern rather than a random mixture of material suspended in fluid. Sanger's success in sequencing insulin spurred on x-ray crystallographers, including Watson and Crick, who by now were trying to understand how DNA directed the formation of proteins within a cell. Soon after attending a series of lectures given by Frederick Sanger in October 1954, Crick began developing a theory which argued that the arrangement of nucleotides in DNA determined the sequence of amino acids in proteins, which in turn helped determine the function of a protein. He published this theory in 1958.

RNA Sequencing

RNA sequencing was one of the earliest forms of nucleotide sequencing. The major landmark of RNA sequencing is the sequence of the first complete gene and the complete genome of Bacteriophage MS2, identified and published by Walter Fiers and his coworkers at the University of Ghent (Ghent, Belgium), in 1972 and 1976. Traditional RNA sequencing methods require the creation of a cDNA molecule which must be sequenced.

Traditional RNA Sequencing Methods

Traditional RNA sequencing methods involve several steps:

- **Reverse Transcription:** The first step is to convert the RNA molecule into a complementary DNA (cDNA) molecule using an enzyme called reverse transcriptase.
- **cDNA Synthesis:** The cDNA molecule is then synthesized through a process called PCR (Polymerase Chain Reaction), which amplifies the cDNA to produce multiple copies.

- **Sequencing:** The amplified cDNA is then sequenced using a technique such as Sanger sequencing or Maxam-Gilbert sequencing.

Challenges and Limitations

Traditional RNA sequencing methods have several limitations. For example:

- They require the creation of a cDNA molecule, which can be time-consuming and labor-intensive. They are prone to errors and biases, which can affect the accuracy of the sequencing results. They are limited in their ability to detect rare or low-abundance transcripts. Advances in RNA Sequencing Technology

In recent years, advances in RNA sequencing technology have addressed some of these limitations. New methods such as next-generation sequencing (NGS) and single-molecule real-timeref>(SMRT) sequencing have enabled faster, more accurate, and more cost-effective sequencing of RNA molecules. These advances have opened up new possibilities for studying gene expression, identifying new genes, and understanding the regulation of gene expression.

Early DNA Sequencing Methods

The first method for determining DNA sequences involved a location-specific primer extension strategy established by Ray Wu at Cornell University in 1970. DNA polymerase catalysis and specific nucleotide labeling were used to sequence the cohesive ends of lambda phage DNA. Between 1970 and 1973, Wu, R Padmanabhan and colleagues demonstrated that this method can be employed to determine any DNA sequence using synthetic location-specific primers. Frederick Sanger then adopted this primer-extension strategy to develop more rapid DNA sequencing methods at the MRC Centre, Cambridge, UK and published a method for "DNA sequencing with chain-terminating inhibitors" in 1977. Walter Gilbert and Allan Maxam at Harvard also developed sequencing methods, including one for "DNA sequencing by chemical degradation". In 1973, Gilbert and Maxam reported the sequence of 24 basepairs using a method known as wandering-spot analysis. Advancements in sequencing were aided by the concurrent development of recombinant DNA technology, allowing DNA samples to be isolated from sources other than viruses.

Sequencing of Full Genomes

The first full DNA genome to be sequenced was that of bacteriophage φX174 in 1977. Medical Research Council scientists deciphered the complete DNA sequence of the Epstein-Barr virus in 1984, finding it contained 172,282 nucleotides. Completion of the sequence marked a significant turning point in DNA sequencing because it was achieved with no prior genetic profile knowledge of the virus.

A non-radioactive method for transferring the DNA molecules of sequencing reaction mixtures onto an immobilizing matrix during electrophoresis was developed by Herbert Pohl and co-workers in the early 1980s. Followed by the commercialization of the DNA sequencer "Direct-Blotting-Electrophoresis-System GATC 1500" by GATC Biotech, which was intensively used in the framework of the EU genome-sequencing programme, the complete DNA sequence of the yeast *Saccharomyces cerevisiae* chromosome II. Leroy E. Hood's laboratory at the California Institute of Technology announced the first semi-automated DNA sequencing machine in 1986. This was followed by Applied Biosystems' marketing of the first fully automated sequencing machine, the ABI 370, in 1987 and by Dupont's Genesis 2000 which used a novel fluorescent labeling technique enabling all four dideoxynucleotides to be identified in a single lane.

By 1990, the U.S. National Institutes of Health (NIH) had begun large-scale sequencing trials on *Mycoplasma capricolum, Escherichia coli, Caenorhabditis elegans*, and *Saccharomyces cerevisiae* at a cost of US$0.75 per base. Meanwhile, sequencing of human cDNA sequences called expressed sequence tags began in Craig Venter's lab, an attempt to capture the coding fraction of the human genome. In 1995, Venter, Hamilton Smith, and colleagues at The Institute for Genomic Research (TIGR) published the first complete genome of a free-living organism, the bacterium *Haemophilus influenzae*. The circular chromosome contains 1,830,137 bases and its publication in the journal Science marked the first published use of whole-genome shotgun sequencing, eliminating the need for initial mapping efforts. By 2001, shotgun sequencing methods had been used to produce a draft sequence of the human genome.

High-Throughput Sequencing (HTS) Methods

Several new methods for DNA sequencing were developed in the mid to late 1990s and were implemented in commercial DNA sequencers by 2000. Together these were called the "next-generation" or "second-generation" sequencing (NGS) methods, in order to distinguish them from the earlier methods, including Sanger sequencing. In contrast to the first generation of sequencing, NGS technology is typically characterized by being highly scalable, allowing the entire genome to be sequenced at once. Usually, this is accomplished by fragmenting the genome into small pieces, randomly sampling for a fragment, and sequencing it using one of a variety of technologies. An entire genome is possible because multiple fragments are sequenced at once (giving it the name "massively parallel" sequencing) in an automated process.

NGS technology has tremendously empowered researchers to look for insights into health, anthropologists to investigate human origins, and is catalyzing the "Personalized Medicine" movement. However, it has also opened the door to more room for error. There are many software tools to carry out the computational analysis of NGS data, often compiled at online platforms such as CSI NGS Portal, each with its own algorithm. Even the parameters within one software package can change the outcome of the analysis. In addition, the large quantities of data produced by DNA sequencing have also required development of new methods and programs for sequence analysis. Several efforts to develop standards in the NGS field have been attempted to address these challenges, most of which have been small-scale efforts arising from individual labs. Most recently, a large, organized, FDA-funded effort has culminated in the BioCompute standard.

On 26 October 1990, Roger Tsien, Pepi Ross, Margaret Fahnestock and Allan J Johnston filed a patent describing stepwise ("base-by-base") sequencing with removable 3' blockers on DNA arrays (blots and single DNA molecules). In 1996, Pål Nyrén and his student Mostafa Ronaghi at the Royal Institute of Technology in Stockholm published their method of pyrosequencing.

On 1 April 1997, Pascal Mayer and Laurent Farinelli submitted patents to the World Intellectual Property Organization describing DNA colony sequencing. The DNA sample preparation and random surface-polymerase chain reaction (PCR) arraying methods described in this patent, coupled to Roger Tsien et al.'s "base-by-base" sequencing method, is now implemented in Illumina's Hi-Seq genome sequencers.

In 1998, Phil Green and Brent Ewing of the University of Washington described their phred quality score for sequencer data analysis, a landmark analysis technique that gained widespread adoption, and which is still the most common metric for assessing the accuracy of a sequencing platform.

Lynx Therapeutics published and marketed massively parallel signature sequencing (MPSS), in 2000. This method incorporated a parallelized, adapter/ligation-mediated, bead-based sequencing technology and served as the first commercially available "next-generation" sequencing method, though no DNA sequencers were sold to independent laboratories.

4.6.4 Basic Methods

Maxam-Gilbert Sequencing

Allan Maxam and Walter Gilbert published a DNA sequencing method in 1977 based on chemical modification of DNA and subsequent cleavage at specific bases. Also known as chemical sequencing, this method allowed purified samples of double-stranded DNA to be used without further cloning. This method's use of radioactive labeling and its technical complexity discouraged extensive use after refinements in the Sanger methods had been made.

Maxam-Gilbert sequencing requires radioactive labeling at one 5' end of the DNA and purification of the DNA fragment to be sequenced. Chemical treatment then generates breaks at a small proportion of one or two of the four nucleotide bases in each of four reactions (G, A+G, C, C+T). The concentration of the modifying chemicals is controlled to introduce on average one modification per DNA molecule. Thus a series of labeled fragments is generated, from the radiolabeled end to the first "cut" site in each molecule. The fragments in the four reactions are electrophoresed side by side in denaturing acrylamide gels for size separation. To visualize the fragments, the gel is exposed to X-ray film for autoradiography, yielding a series of dark bands each corresponding to a radiolabeled DNA fragment, from which the sequence may be inferred.

Chain-Termination Methods

The chain-termination method developed by Frederick Sanger and coworkers in 1977 soon became the method of choice, owing to its relative ease and reliability. When invented, the chain-terminator method used fewer toxic chemicals and lower amounts of radioactivity than the Maxam and Gilbert method. Because of its comparative ease, the Sanger method was soon automated and was the method used in the first generation of DNA sequencers.

Sanger sequencing is the method which prevailed from the 1980s until the mid-2000s. Over that period, great advances were made in the technique, such as fluorescent labelling, capillary electrophoresis, and general automation. These developments allowed much more efficient sequencing, leading to lower costs. The Sanger method, in mass production form, is the technology which produced the first human genome in 2001, ushering in the age of genomics. However, later in the decade, radically different approaches reached the market, bringing the cost per genome down from $100 million in 2001 to $10,000 in 2011.

Sequencing by Synthesis

The objective for sequential sequencing by synthesis (SBS) is to determine the sequencing of a DNA sample by detecting the incorporation of a nucleotide by a DNA polymerase. An engineered polymerase is used to synthesize a copy of a single strand of DNA and the incorporation of each nucleotide is monitored. The principle of real-time sequencing by synthesis was first described in 1993 with improvements published some years later. The key parts are highly similar for all embodiments of SBS and includes (1) amplification of DNA (to enhance the subsequent signal) and attach the DNA to be sequenced to a solid support, (2) generation of single stranded DNA on the solid support, (3) incorporation of nucleotides using an engineered polymerase and (4) real-time detection of the incorporation of nucleotide The steps 3-4 are repeated and the sequence is assembled from the signals obtained in step 4. This principle of real-time sequencing-by-synthesis has been used for almost all massive parallel sequencing instruments, including 454, PacBio, IonTorrent, Illumina and MGI.

4.6.5 Large-Scale Sequencing and de novo Sequencing

Large-scale sequencing often aims at sequencing very long DNA pieces, such as whole chromosomes, although large-scale sequencing can also be used to generate very large numbers of short sequences, such as found in phage display. For longer targets such as chromosomes, common approaches consist of cutting (with restriction enzymes) or shearing (with mechanical forces) large DNA fragments into shorter DNA fragments. The fragmented DNA may then be cloned into a DNA vector and amplified in a bacterial host such as *Escherichia coli*. Short DNA fragments purified from individual bacterial colonies are individually sequenced and assembled electronically into one long, contiguous sequence. Studies have shown that adding a size selection step to collect DNA fragments of uniform size can improve sequencing efficiency and accuracy of the genome assembly. In these studies, automated sizing has proven to be more reproducible and precise than manual gel sizing.

The term "*de novo* sequencing" specifically refers to methods used to determine the sequence of DNA with no previously known sequence. *De novo* translates from Latin as "from the beginning". Gaps in the assembled sequence may be filled by primer walking. The different strategies have different tradeoffs in speed and accuracy; shotgun methods are often used for sequencing large genomes, but its assembly is complex and difficult, particularly with sequence repeats often causing gaps in genome assembly.

Most sequencing approaches use an *in vitro* cloning step to amplify individual DNA molecules, because their molecular detection methods are not sensitive enough for single molecule sequencing. Emulsion PCR isolates individual DNA molecules along with primer-coated beads in aqueous droplets within an oil phase. A polymerase chain reaction (PCR) then coats each bead with clonal copies of the DNA molecule followed by immobilization for later sequencing. Emulsion PCR is used in the methods developed by Marguilis et al. (commercialized by 454 Life Sciences), Shendure and Porreca et al. (also known as "polony sequencing") and SOLiD sequencing, (developed by Agencourt, later Applied Biosystems, now Life Technologies). Emulsion PCR is also used in the GemCode and Chromium platforms developed by 10x Genomics.

Shotgun Sequencing

Shotgun sequencing is a sequencing method designed for analysis of DNA sequences longer than 1000 base pairs, up to and including entire chromosomes. This method requires the target DNA to be broken into random fragments. After sequencing individual fragments using the chain termination method, the sequences can be reassembled on the basis of their overlapping regions.

4.6.6 High-Throughput Methods

High-throughput sequencing, which includes next-generation "short-read" and third-generation "long-read" sequencing methods, applies to exome sequencing, genome sequencing, genome resequencing, transcriptome profiling (RNA-Seq), DNA-protein interactions (ChIP-sequencing), and epigenome characterization.

Long-Read Sequencing Methods

Single Molecule Real Time (SMRT) Sequencing

SMRT sequencing is based on the sequencing by synthesis approach. The DNA is synthesized in zero-mode wave-guides (ZMWs) – small well-like containers with the capturing tools located at the bottom of the well. The sequencing is performed with use of unmodified polymerase (attached to the ZMW bottom) and fluorescently labelled nucleotides flowing freely in the solution. The wells are constructed in a way that only the fluorescence occurring by the bottom of the well is detected. The fluorescent label is detached from the nucleotide upon its incorporation into the DNA strand, leaving an unmodified DNA strand. According to Pacific Biosciences (PacBio), the SMRT technology developer, this methodology allows detection of nucleotide modifications (such as cytosine methylation). This happens through the observation of polymerase kinetics. This approach allows reads of 20,000 nucleotides or more, with average read lengths of 5 kilobases. In 2015, Pacific Biosciences announced the launch of a new sequencing instrument called the Sequel System, with 1 million ZMWs compared to 150,000 ZMWs in the PacBio RS II instrument. SMRT sequencing is referred to as "third-generation" or "long-read" sequencing.

Nanopore DNA Sequencing

The DNA passing through the nanopore changes its ion current. This change is dependent on the shape, size and length of the DNA sequence. Each type of the nucleotide blocks the ion flow through the pore for a different period of time. The method does not require modified nucleotides and is performed in real time. Nanopore sequencing is referred to as "third-generation" or "long-read" sequencing, along with SMRT sequencing.

Early industrial research into this method was based on a technique called 'exonuclease sequencing', where the readout of electrical signals occurred as nucleotides passed by alpha(α)-hemolysin pores covalently bound with cyclodextrin. However the subsequent commercial method, ‹strand sequencing›, sequenced DNA bases in an intact strand.

Two main areas of nanopore sequencing in development are solid state nanopore sequencing, and protein based nanopore sequencing. Protein nanopore sequencing utilizes membrane protein complexes such as α-hemolysin, MspA (*Mycobacterium smegmatis* Porin A) or CssG, which show great promise given their ability to distinguish between individual and groups of nucleotides. In contrast, solid-state nanopore sequencing utilizes synthetic materials such as silicon nitride and aluminum oxide and it is preferred for its superior mechanical ability and thermal and chemical stability. The fabrication method is essential for this type of sequencing given that the nanopore array can contain hundreds of pores with diameters smaller than eight nanometers.

The concept originated from the idea that single stranded DNA or RNA molecules can be electrophoretically driven in a strict linear sequence through a biological pore that can be less than eight nanometers, and can be detected given that the molecules release an ionic current while moving through the pore. The pore contains a detection region capable of recognizing different bases, with each base generating various time specific signals corresponding to the sequence of bases as they cross the pore which are then evaluated. Precise control over the DNA transport through the pore is crucial for success. Various enzymes such as exonucleases and polymerases have been used to moderate this process by positioning them near the pore's entrance.

Short-Read Sequencing Methods

Massively Parallel Signature Sequencing (MPSS)

The first of the high-throughput sequencing technologies, massively parallel signature sequencing (or MPSS, also called next generation sequencing), was developed in the 1990s at Lynx Therapeutics, a company founded in 1992 by Sydney Brenner and Sam Eletr. MPSS was a bead-based method that used a complex approach of adapter ligation followed by adapter decoding, reading the sequence in increments of four nucleotides. This method made it susceptible to sequence-specific bias or loss of specific sequences. Because the technology was so complex, MPSS was only performed 'in-house' by Lynx Therapeutics and no DNA sequencing machines were sold to independent laboratories. Lynx Therapeutics merged with Solexa (later acquired by Illumina) in 2004, leading to the development of sequencing-by-synthesis, a simpler approach acquired from Manteia Predictive Medicine, which rendered MPSS obsolete. However, the essential properties of the MPSS output were typical of later high-throughput data types, including hundreds of thousands of short DNA sequences. In the case of MPSS, these were typically used for sequencing cDNA for measurements of gene expression levels.

Polony Sequencing

The polony sequencing method, developed in the laboratory of George M. Church at Harvard, was among the first high-throughput sequencing systems and was used to sequence a full *E. coli* genome in 2005. It combined an in vitro paired-tag library with emulsion PCR, an automated microscope, and ligation-based sequencing chemistry to sequence an *E. coli* genome at an accuracy of >99.9999% and a cost approximately 1/9 that of Sanger sequencing. The technology was licensed to Agencourt Biosciences, subsequently spun out into Agencourt Personal Genomics, and eventually incorporated into the Applied Biosystems SOLiD platform. Applied Biosystems was later acquired by Life Technologies, now part of Thermo Fisher Scientific.

454 Pyrosequencing

A parallelized version of pyrosequencing was developed by 454 Life Sciences, which has since been acquired by Roche Diagnostics. The method amplifies DNA inside water droplets in an oil solution (emulsion PCR), with each droplet containing a single DNA template attached to a single primer-coated bead that then forms a clonal colony. The sequencing machine contains many picoliter-volume wells each containing a single bead and sequencing enzymes. Pyrosequencing uses luciferase to generate light for detection of the individual nucleotides added to the nascent DNA, and the combined data are used to generate sequence reads. This technology provides intermediate read length and price per base compared to Sanger sequencing on one end and Solexa and SOLiD on the other.

Illumina (Solexa) Sequencing

Solexa, now part of Illumina, was founded by Shankar Balasubramanian and David Klenerman in 1998, and developed a sequencing method based on reversible dye-terminators technology, and engineered polymerases. The reversible terminated chemistry concept was invented by Bruno Canard and Simon Sarfati at the Pasteur Institute in Paris. It was developed internally at Solexa by those named on the relevant patents. In 2004, Solexa acquired the company Manteia Predictive Medicine in order to gain a massively parallel sequencing technology invented in 1997 by Pascal Mayer and Laurent Farinelli. It is based on "DNA clusters" or "DNA colonies", which involves the clonal amplification of DNA on a surface. The cluster technology was co-acquired with Lynx Therapeutics of California. Solexa Ltd. later merged with Lynx to form Solexa Inc.

In this method, DNA molecules and primers are first attached on a slide or flow cell and amplified with polymerase so that local clonal DNA colonies, later coined "DNA clusters", are formed. To determine the sequence, four types of reversible terminator bases (RT-bases) are added and non-incorporated nucleotides are washed away. A camera takes images of the fluorescently labeled nucleotides. Then the dye, along with the terminal 3' blocker, is chemically removed from the DNA, allowing for the next cycle to begin. Unlike pyrosequencing, the DNA chains are extended one nucleotide at a time and image acquisition can be performed at a delayed moment, allowing for very large arrays of DNA colonies to be captured by sequential images taken from a single camera.

Decoupling the enzymatic reaction and the image capture allows for optimal throughput and theoretically unlimited sequencing capacity. With an optimal configuration, the ultimately reachable instrument throughput is thus dictated solely by the analog-to-digital conversion rate of the camera, multiplied by the number of cameras and divided by the number of pixels per DNA colony required for visualizing them optimally (approximately 10 pixels/colony). In 2012, with cameras operating at more than 10 MHz A/D conversion rates and available optics, fluidics and enzymatics, throughput can be multiples of 1 million nucleotides/second, corresponding roughly to 1 human genome equivalent at 1x coverage per hour per instrument, and 1 human genome re-sequenced (at approx. 30x) per day per instrument (equipped with a single camera).

Combinatorial Probe Anchor Synthesis (cPAS)

This method is an upgraded modification to combinatorial probe anchor ligation technology (cPAL) described by Complete Genomics which has since become part of Chinese genomics company BGI in 2013. The two companies have refined the technology to allow for longer read lengths, reaction time reductions and faster time to results. In addition, data are now generated as contiguous full-length reads in the standard FASTQ file format and can be used as-is in most short-read-based bioinformatics analysis pipelines.

The two technologies that form the basis for this high-throughput sequencing technology are DNA nanoballs (DNB) and patterned arrays for nanoball attachment to a solid surface. DNA nanoballs are simply formed by denaturing double stranded, adapter ligated libraries and ligating the forward strand only to a splint oligonucleotide to form a ssDNA circle. Faithful copies of the circles containing the DNA insert are produced utilizing Rolling Circle Amplification that generates approximately 300–500 copies. The long strand of ssDNA folds upon itself to produce a three-dimensional nanoball structure that is approximately 220 nm in diameter. Making DNBs replaces the need to generate PCR copies of the library on the flow cell and as such can remove large proportions of duplicate reads, adapter-adapter ligations and PCR induced errors.

The patterned array of positively charged spots is fabricated through photolithography and etching techniques followed by chemical modification to generate a sequencing flow cell. Each spot on the flow cell is approximately 250 nm in diameter, are separated by 700 nm (centre to centre) and allows easy attachment of a single negatively charged DNB to the flow cell and thus reducing under or over-clustering on the flow cell.

Sequencing is then performed by addition of an oligonucleotide probe that attaches in combination to specific sites within the DNB. The probe acts as an anchor that then allows one of four single reversibly inactivated, labelled nucleotides to bind after flowing across the flow cell. Unbound nucleotides are washed away before laser excitation of the attached labels then emit fluorescence and signal is captured by cameras that is converted to a digital output for base calling. The attached base has its terminator and label chemically cleaved at completion of the cycle. The cycle is repeated with another flow of free, labelled nucleotides across the flow cell to allow the next nucleotide to bind and have its signal captured. This process is completed a number of times (usually 50 to 300 times) to determine the sequence of the inserted piece of DNA at a rate of approximately 40 million nucleotides per second as of 2018.

SOLiD Sequencing

Applied Biosystems' (now a Life Technologies brand) SOLiD technology employs sequencing by ligation. Here, a pool of all possible oligonucleotides of a fixed length are labeled according to the sequenced position. Oligonucleotides are annealed and ligated; the preferential ligation by DNA ligase for matching sequences results in a signal informative of the nucleotide at that position. Each base in the template is sequenced twice, and the resulting data are decoded according to the 2 base encoding scheme used in this method. Before sequencing, the DNA is amplified by emulsion PCR. The resulting beads, each containing single copies of the same DNA molecule, are deposited on a glass slide. The result is sequences of quantities and lengths comparable to Illumina sequencing. This sequencing by ligation method has been reported to have some issue sequencing palindromic sequences.

Ion Torrent Semiconductor Sequencing

Ion Torrent Systems Inc. (now owned by Life Technologies) developed a system based on using standard sequencing chemistry, but with a novel, semiconductor-based detection system. This method of sequencing is based on the detection of hydrogen ions that are released during the polymerisation of DNA, as opposed to the optical methods used in other sequencing systems. A microwell containing a template DNA strand to be sequenced is flooded with a single type of nucleotide. If the introduced nucleotide is complementary to the leading template nucleotide it is incorporated into the growing complementary strand. This causes the release of a hydrogen ion that triggers a hypersensitive ion sensor, which indicates that a reaction has occurred. If homopolymer repeats are present in the template sequence, multiple nucleotides will be incorporated in a single cycle. This leads to a corresponding number of released hydrogens and a proportionally higher electronic signal.

DNA Nanoball Sequencing

DNA nanoball sequencing is a type of high throughput sequencing technology used to determine the entire genomic sequence of an organism. The company Complete Genomics uses this technology to sequence samples submitted by independent researchers. The method uses rolling circle replication to amplify small fragments of genomic DNA into DNA nanoballs. Unchained sequencing by ligation is then used to determine the nucleotide sequence. This method of DNA sequencing allows large numbers of DNA nanoballs to be sequenced per run and at low reagent costs compared to other high-throughput sequencing platforms. However, only short sequences of DNA are determined from each DNA nanoball which makes mapping the short reads to a reference genome difficult.

Heliscope Single Molecule Sequencing

Heliscope sequencing is a method of single-molecule sequencing developed by Helicos Biosciences. It uses DNA fragments with added poly-A tail adapters which are attached to the flow cell surface. The next steps involve extension-based sequencing with cyclic washes of the flow cell with fluorescently labeled nucleotides (one nucleotide type at a time, as with the Sanger method). The reads are performed by the Heliscope sequencer. The reads are short, averaging 35 bp. What made this technology especially novel was that it was the first of its class to sequence non-amplified DNA, thus preventing any read errors associated with amplification steps. In 2009 a human genome was sequenced using the Heliscope, however in 2012 the company went bankrupt.

Microfluidic Systems

There are two main microfluidic systems that are used to sequence DNA; droplet based microfluidics and digital microfluidics. Microfluidic devices solve many of the current limitations of current sequencing arrays.

Abate et al. studied the use of droplet-based microfluidic devices for DNA sequencing. These devices have the ability to form and process picoliter sized droplets at the rate of thousands per second. The devices were created from polydimethylsiloxane (PDMS) and used Forster resonance energy transfer, FRET assays to read the sequences of DNA encompassed in the droplets. Each position on the array tested for a specific 15 base sequence.

Fair et al. used digital microfluidic devices to study DNA pyrosequencing. Significant advantages include the portability of the device, reagent volume, speed of analysis, mass manufacturing abilities, and high throughput. This study provided a proof of concept showing that digital devices can be used for pyrosequencing; the study included using synthesis, which involves the extension of the enzymes and addition of labeled nucleotides.

Boles et al. also studied pyrosequencing on digital microfluidic devices. They used an electro-wetting device to create, mix, and split droplets. The sequencing uses a three-enzyme protocol and DNA templates anchored with magnetic beads. The device was tested using two protocols and resulted in 100% accuracy based on raw pyrogram levels. The advantages of these digital microfluidic devices include size, cost, and achievable levels of functional integration.

DNA sequencing research, using microfluidics, also has the ability to be applied to the sequencing of RNA, using similar droplet microfluidic techniques, such as the method, inDrops. This shows that many of these DNA sequencing techniques will be able to be applied further and be used to understand more about genomes and transcriptomes.

4.6.7 Methods in Development

DNA sequencing methods currently under development include reading the sequence as a DNA strand transits through nanopores (a method that is now commercial but subsequent generations such as solid-state nanopores are still in development), and microscopy-based techniques, such as atomic force microscopy or transmission electron microscopy that are used to identify the positions of individual nucleotides within long DNA fragments (>5,000 bp) by nucleotide labeling with heavier elements (e.g., halogens) for visual detection and recording. Third generation technologies aim to increase throughput and decrease the time to result and cost by eliminating the need for excessive reagents and harnessing the processivity of DNA polymerase.

Tunnelling Currents DNA Sequencing

Another approach uses measurements of the electrical tunnelling currents across single-strand DNA as it moves through a channel. Depending on its electronic structure, each base affects the tunnelling current differently, allowing differentiation between different bases. The use of tunnelling currents has the potential to sequence orders of magnitude faster than ionic current methods and the sequencing of several DNA oligomers and micro-RNA has already been achieved.

Sequencing by Hybridization

Sequencing by hybridization is a non-enzymatic method that uses a DNA microarray. A single pool of DNA whose sequence is to be determined is fluorescently labeled and hybridized to an array containing known sequences. Strong hybridization signals from a given spot on the array identifies its sequence in the DNA being sequenced. This method of sequencing utilizes binding characteristics of a library of short single stranded DNA molecules (oligonucleotides), also called DNA probes, to reconstruct a target DNA sequence. Non-specific hybrids are removed by washing and the target DNA is eluted. Hybrids are re-arranged such that the DNA sequence can be reconstructed. The benefit of this sequencing type is its ability to capture a large number of targets with a homogenous coverage. A large number of chemicals and starting DNA is usually required. However, with the advent of solution-based hybridization, much less equipment and chemicals are necessary.

Sequencing with Mass Spectrometry

Mass spectrometry may be used to determine DNA sequences. Matrix-assisted laser desorption ionization time-of-flight mass spectrometry, or MALDI-TOF MS, has specifically been investigated as an alternative method to gel electrophoresis for visualizing DNA fragments. With this method, DNA fragments generated by chain-termination sequencing reactions are compared by mass rather than by size. The mass of each nucleotide is different from the others and this difference is detectable by mass spectrometry. Single-nucleotide mutations in a fragment can be more easily detected with MS than by gel electrophoresis alone. MALDI-TOF MS can more easily detect differences between RNA fragments, so researchers may indirectly sequence DNA with MS-based methods by converting it to RNA first.

The higher resolution of DNA fragments permitted by MS-based methods is of special interest to researchers in forensic science, as they may wish to find single-nucleotide polymorphisms in human DNA samples to identify individuals. These samples may be highly degraded so forensic researchers often prefer mitochondrial DNA for its higher stability and applications for lineage studies. MS-based sequencing methods have been used to compare the sequences of human mitochondrial DNA from samples in a Federal Bureau of Investigation database and from bones found in mass graves of World War I soldiers.

Early chain-termination and TOF MS methods demonstrated read lengths of up to 100 base pairs. Researchers have been unable to exceed this average read size; like chain-termination sequencing alone, MS-based DNA sequencing may not be suitable for large *de novo* sequencing projects. Even so, a recent study did use the short sequence reads and mass spectroscopy to compare single-nucleotide polymorphisms in pathogenic *Streptococcus* strains.

Microfluidic Sanger Sequencing

In microfluidic Sanger sequencing the entire thermocycling amplification of DNA fragments as well as their separation by electrophoresis is done on a single glass wafer (approximately 10 cm in diameter) thus reducing the reagent usage as well as cost. In some instances researchers have shown that they can increase the throughput of conventional sequencing through the use of microchips. Research will still need to be done in order to make this use of technology effective.

Microscopy-Based Techniques

This approach directly visualizes the sequence of DNA molecules using electron microscopy. The first identification of DNA base pairs within intact DNA molecules by enzymatically incorporating modified bases, which contain atoms of increased atomic number, direct visualization and identification of individually labeled bases within a synthetic 3,272 base-pair DNA molecule and a 7,249 base-pair viral genome has been demonstrated.

RNAP Sequencing

This method is based on use of RNA polymerase (RNAP), which is attached to a polystyrene bead. One end of DNA to be sequenced is attached to another bead, with both beads being placed in optical traps. RNAP motion during transcription brings the beads in closer and their relative distance changes, which can then be recorded at a single nucleotide resolution. The sequence is deduced based on the four readouts with lowered concentrations of each of the four nucleotide types, similarly to the Sanger method. A comparison is made between regions and sequence information is deduced by comparing the known sequence regions to the unknown sequence regions.

In Vitro Virus High-Throughput Sequencing

A method has been developed to analyze full sets of protein interactions using a combination of 454 pyrosequencing and an *in vitro* virus mRNA display method. Specifically, this method covalently links proteins of interest to the mRNAs encoding them, then detects the mRNA pieces using reverse transcription PCRs. The mRNA may then be amplified and sequenced. The combined method was titled IVV-HiTSeq and can be performed under cell-free conditions, though its results may not be representative of *in vivo* conditions.

4.6.8 Sample Preparation

The success of any DNA sequencing protocol relies upon the DNA or RNA sample extraction and preparation from the biological material of interest.

- A successful DNA extraction will yield a DNA sample with long, non-degraded strands.
- A successful RNA extraction will yield a RNA sample that should be converted to complementary DNA (cDNA) using reverse transcriptase—a DNA polymerase that synthesizes a complementary DNA based on existing strands of RNA in a PCR-like manner. Complementary DNA can then be processed the same way as genomic DNA.

After DNA or RNA extraction, samples may require further preparation depending on the sequencing method. For Sanger sequencing, either cloning procedures or

PCR are required prior to sequencing. In the case of next-generation sequencing methods, library preparation is required before processing. Assessing the quality and quantity of nucleic acids both after extraction and after library preparation identifies degraded, fragmented, and low-purity samples and yields high-quality sequencing data.

4.6.9 Development Initiatives

In October 2006, the X Prize Foundation established an initiative to promote the development of full genome sequencing technologies, called the Archon X Prize, intending to award $10 million to "the first Team that can build a device and use it to sequence 100 human genomes within 10 days or less, with an accuracy of no more than one error in every 100,000 bases sequenced, with sequences accurately covering at least 98% of the genome, and at a recurring cost of no more than $10,000 (US) per genome." Each year the National Human Genome Research Institute, or NHGRI, promotes grants for new research and developments in genomics. 2010 grants and 2011 candidates include continuing work in microfluidic, polony and base-heavy sequencing methodologies.

4.6.10 Computational Challenges

The sequencing technologies described here produce raw data that needs to be assembled into longer sequences such as complete genomes (sequence assembly). There are many computational challenges to achieve this, such as the evaluation of the raw sequence data which is done by programs and algorithms such as Phred and Phrap. Other challenges have to deal with repetitive sequences that often prevent complete genome assemblies because they occur in many places of the genome. As a consequence, many sequences may not be assigned to particular chromosomes. The production of raw sequence data is only the beginning of its detailed bioinformatical analysis. Yet new methods for sequencing and correcting sequencing errors were developed.

4.6.11 Ethical Issues

Human genetics have been included within the field of bioethics since the early 1970s and the growth in the use of DNA sequencing (particularly high-throughput sequencing) has introduced a number of ethical issues. One key issue is the ownership of an individual's DNA and the data produced when that DNA is sequenced.

Regarding the DNA molecule itself, the leading legal case on this topic, *Moore v. Regents of the University of California* (1990) ruled that individuals have no property rights to discarded cells or any profits made using these cells (for instance, as a patented cell line). However, individuals have a right to informed consent regarding removal and use of cells. Regarding the data produced through DNA sequencing, *Moore* gives the individual no rights to the information derived from their DNA.

As DNA sequencing becomes more widespread, the storage, security and sharing of genomic data has also become more important. For instance, one concern is that insurers may use an individual's genomic data to modify their quote, depending on the perceived future health of the individual based on their DNA. In May 2008, the Genetic Information Nondiscrimination Act (GINA) was signed in the United States, prohibiting discrimination on the basis of genetic information with respect to health insurance and employment.

In 2012, the US Presidential Commission for the Study of Bioethical Issues reported that existing privacy legislation for DNA sequencing data such as GINA and the Health Insurance Portability and Accountability Act were insufficient, noting that whole-genome sequencing data was particularly sensitive, as it could be used to identify not only the individual from which the data was created, but also their relatives.

In most of the United States, DNA that is "abandoned", such as that found on a licked stamp or envelope, coffee cup, cigarette, chewing gum, household trash, or hair that has fallen on a public sidewalk, may legally be collected and sequenced by anyone, including the police, private investigators, political opponents, or people involved in paternity disputes. As of 2013, eleven states have laws that can be interpreted to prohibit "DNA theft".

Ethical issues have also been raised by the increasing use of genetic variation screening, both in newborns, and in adults by companies such as 23andMe. It has been asserted that screening for genetic variations can be harmful, increasing anxiety in individuals who have been found to have an increased risk of disease.

For example, in one case noted in *Time*, doctors screening an ill baby for genetic variants chose not to inform the parents of an unrelated variant linked to dementia due to the harm it would cause to the parents. However, a 2011 study in *The New England Journal of Medicine* has shown that individuals undergoing disease risk profiling did not show increased levels of anxiety. Also, the development of Next Generation sequencing technologies such as Nanopore based sequencing has also raised further ethical concerns.

4.7 HOST CELLS: FACTORIES FOR DNA CLONING AND PROTEIN PRODUCTION

Host cells, particularly bacterial cells such as *Escherichia coli*, are used to replicate recombinant DNA and express proteins. After recombinant DNA is introduced into the host cell, the cell's machinery replicates the DNA and can transcribe and translate it into protein if necessary. Host cells are chosen based on the requirements of the experiment, with bacterial cells commonly used for basic cloning and expression of non-complex proteins.

In addition to bacteria, other host systems, such as yeast, insect, and mammalian cells, are also used, particularly when complex eukaryotic proteins are required. Each host system has unique advantages and limitations, and choosing the appropriate host cell is essential for obtaining the desired gene product or protein.

4.8 HYBRIDIZATION TECHNIQUES: DETECTING SPECIFIC DNA SEQUENCES

Hybridization techniques, such as Southern blotting, Northern blotting, and in situ hybridization, are methods used to detect specific DNA or RNA sequences. These methods rely on the principle of complementary base pairing, where a labeled DNA or RNA probe binds to a target sequence in a sample. Southern blotting detects specific DNA sequences, while Northern blotting is used for RNA, and in situ hybridization allows detection of target sequences within intact cells or tissues.

These techniques are crucial for confirming the presence and location of specific DNA or RNA sequences, analyzing gene expression patterns, and identifying mutations. Hybridization has been instrumental in genetic research, diagnostics, and gene mapping, providing a versatile tool for recombinant DNA studies.

In biology, a hybrid is the offspring resulting from combining the qualities of two organisms of different varieties, subspecies, species or genera through sexual reproduction. Generally, it means that each cell has genetic material from two different organisms, whereas an individual where some cells are derived from a different organism is called a chimera. Hybrids are not always intermediates between their parents such as in blending inheritance (a now discredited theory in modern genetics by particulate inheritance), but can show hybrid vigor, sometimes growing larger or taller than either parent. The concept of a hybrid is interpreted differently in animal and plant breeding, where there is interest in the individual parentage. In genetics, attention is focused on the numbers of chromosomes. In taxonomy, a key question is how closely related the parent species are.

Species are reproductively isolated by strong barriers to hybridization, which include genetic and morphological differences, differing times of fertility, mating behaviors and cues, and physiological rejection of sperm cells or the developing embryo. Some act before fertilization and others after it. Similar barriers exist in plants, with differences in flowering times, pollen vectors, inhibition of pollen tube growth, somatoplastic sterility, cytoplasmic-genic male sterility and the structure of the chromosomes. A few animal species and many plant species, however, are the result of hybrid speciation, including important crop plants such as wheat, where the number of chromosomes has been doubled.

A form of often intentional human-mediated hybridization is the crossing of wild and domesticated species. This is common in both traditional horticulture and modern agriculture; many commercially useful fruits, flowers, garden herbs, and trees have been produced by hybridization. One such flower, *Oenothera lamarckiana*, was central to early genetics research into mutationism and polyploidy. It is also more occasionally done in the livestock and pet trades; some well-known wild × domestic hybrids are beefalo and wolfdogs. Human selective breeding of domesticated animals and plants has also resulted in the development of distinct breeds (usually called cultivars in reference to plants); crossbreeds between them (without any wild stock) are sometimes also imprecisely referred to as "hybrids".

Hybrid humans existed in prehistory. For example, Neanderthals and anatomically modern humans are thought to have interbred as recently as 40,000 years ago.

Mythological hybrids appear in human culture in forms as diverse as the Minotaur, blends of animals, humans and mythical beasts such as centaurs and sphinxes, and the Nephilim of the Biblical apocrypha described as the wicked sons of fallen angels and attractive women.

4.8.1 Significance

In Evolution

Hybridization between species plays an important role in evolution, though there is much debate about its significance. Roughly 25% of plants and 10% of animals are known to form hybrids with at least one other species. One example of an adaptive benefit to hybridization is that hybrid individuals can form a "bridge" transmitting potentially helpful genes from one species to another when the hybrid backcrosses with one of its parent species, a process called introgression. Hybrids can also cause speciation, either because the hybrids are genetically incompatible with their parents and not each other, or because the hybrids occupy a different niche than either parent.

Hybridization is a particularly common mechanism for speciation in plants, and is now known to be fundamental to the evolutionary history of plants. Plants frequently form polyploids, individuals with more than two copies of each chromosome. Whole genome doubling has occurred repeatedly in plant evolution. When two plant species hybridize, the hybrid may double its chromosome count by incorporating the entire nuclear genome of both parents, resulting in offspring that are reproductively incompatible with either parent because of different chromosome counts.

In Conservation

Human impact on the environment has resulted in an increase in the interbreeding between regional species, and the proliferation of introduced species worldwide has also resulted in an increase in hybridization. This has been referred to as genetic pollution out of concern that it may threaten many species with extinction. Similarly, genetic erosion from monoculture in crop plants may be damaging the gene pools of many species for future breeding.

The conservation impacts of hybridization between species are highly debated. While hybridization could potentially threaten rare species or lineages by "swamping" the genetically "pure" individuals with hybrids, hybridization could also save a rare lineage from extinction by introducing genetic diversity. It has been proposed that hybridization could be a useful tool to conserve biodiversity by allowing organisms to adapt, and that efforts to preserve the separateness of a "pure" lineage could harm conservation by lowering the organisms' genetic diversity and adaptive potential, particularly in species with low populations. While endangered species are often protected by law, hybrids are often excluded from protection, resulting in challenges to conservation.

4.8.2 As Seen by Different Disciplines

Animal and Plant Breeding

From the point of view of animal and plant breeders, there are several kinds of hybrid formed from crosses within a species, such as between different breeds. Single cross hybrids result from the cross between two true-breeding organisms which produces an F1 hybrid (first filial generation). The cross between two different homozygous lines produces an F1 hybrid that is heterozygous; having two alleles, one contributed by each parent and typically one is dominant and the other recessive. Typically, the F1 generation is also phenotypically homogeneous, producing offspring that are all similar to each other. Double cross hybrids result from the cross between two different F1 hybrids (i.e., there are four unrelated grandparents). Three-way cross hybrids result from the cross between an F1 hybrid and an inbred line. Triple cross hybrids result from the crossing of two different three-way cross hybrids. Top cross (or "topcross") hybrids result from the crossing of a top quality or pure-bred male and a lower quality female, intended to improve the quality of the offspring, on average.

Population hybrids result from the crossing of plants or animals in one population with those of another population. These include interspecific hybrids or crosses between different breeds. In biology, the result of crossing of two populations is called a synthetic population.

In horticulture, the term stable hybrid is used to describe an annual plant that, if grown and bred in a small monoculture free of external pollen (e.g., an air-filtered greenhouse) produces offspring that are "true to type" with respect to phenotype; i.e., a true-breeding organism.

Biogeography

Hybridization can occur in the hybrid zones where the geographical ranges of species, subspecies, or distinct genetic lineages overlap. For example, the butterfly *Limenitis arthemis* has two major subspecies in North America, *L. a. arthemis* (the white admiral) and *L. a. astyanax* (the red-spotted purple). The white admiral has a bright, white band on its wings, while the red-spotted purple has cooler blue-green shades. Hybridization occurs between a narrow area across New England, southern Ontario, and the Great Lakes, the "suture region". It is at these regions that the subspecies were formed. Other hybrid zones have formed between described species of plants and animals.

Genetics

From the point of view of genetics, several different kinds of hybrid can be distinguished. A genetic hybrid carries two different alleles of the same gene, where for instance one allele may code for a lighter coat colour than the other. A structural hybrid results from the fusion of gametes that have differing structure in at least one chromosome, as a result of structural abnormalities. A numerical hybrid results from the fusion of gametes having different haploid numbers of chromosomes. A permanent hybrid results when only the heterozygous genotype occurs, as in *Oenothera lamarckiana*, because all homozygous combinations are lethal. In the early history of genetics, Hugo de Vries supposed these were caused by mutation.

Genetic Complementation

Genetic complementation is a hybridization test widely used in genetics to determine whether two separately isolated mutants that have the same (or similar) phenotype are defective in the same gene or in different genes. If a hybrid organism containing the genomes of two different mutant parental organisms displays a wild type phenotype, it is ordinarily considered that the two parental mutant organisms are defective in different genes. If the hybrid organism displays a distinctly mutant phenotype, the two mutant parental organisms are considered to be defective in the same gene. However, in some cases the hybrid organism may display a phenotype that is only weakly (or partially) wild-type, and this may reflect intragenic (interallelic) complementation.

Taxonomy

From the point of view of taxonomy, hybrids differ according to their parentage. Hybrids between different subspecies (such as between the dog and Eurasian wolf) are called intra-specific hybrids. Interspecific hybrids are the offspring from interspecies mating; these sometimes result in hybrid speciation. Intergeneric hybrids result from matings between different genera, such as between sheep and goats. Interfamilial hybrids, such as between chickens and guineafowl or pheasants, are reliably described but extremely rare. Interordinal hybrids (between different orders) are few, but have been engineered between the sea urchin *Strongylocentrotus purpuratus* (female) and the sand dollar *Dendraster excentricus* (male).

4.8.3 Biology

Expression of Parental Traits

When two distinct types of organisms breed with each other, the resulting hybrids typically have intermediate traits (e.g., one plant parent has red flowers, the other has white, and the hybrid, pink flowers). Commonly, hybrids also combine traits seen only separately in one parent or the other (e.g., a bird hybrid might combine the yellow head of one parent with the orange belly of the other).

Mechanisms of Reproductive Isolation

Interspecific hybrids are bred by mating individuals from two species, normally from within the same genus. The offspring display traits and characteristics of both parents, but are often sterile, preventing gene flow between the species. Sterility is often attributed to the different number of chromosomes between the two species. For example, donkeys have 62 chromosomes, horses have 64 chromosomes, and mules or hinnies have 63 chromosomes.

Mules, hinnies, and other normally sterile interspecific hybrids cannot produce viable gametes, because differences in chromosome structure prevent appropriate pairing and segregation during meiosis, meiosis is disrupted, and viable sperm and eggs are not formed. However, fertility in female mules has been reported with a donkey as the father.

A variety of mechanisms limit the success of hybridization, including the large genetic difference between most species. Barriers include morphological differences, differing times of fertility, mating behaviors and cues, and physiological rejection of sperm cells or the developing embryo. Some act before fertilization; others after it.

In plants, some barriers to hybridization include blooming period differences, different pollinator vectors, inhibition of pollen tube growth, somatoplastic sterility, cytoplasmic-genic male sterility and structural differences of the chromosomes.

Speciation

A few animal species are the result of hybridization. The Lonicera fly is a natural hybrid. The American red wolf appears to be a hybrid of the gray wolf and the coyote, although its taxonomic status has been a subject of controversy. The European edible frog is a semi-permanent hybrid between pool frogs and marsh frogs; its population requires the continued presence of at least one of the parent species. Cave paintings indicate that the European bison is a natural hybrid of the aurochs and the steppe bison.

Plant hybridization is more commonplace compared to animal hybridization. Many crop species are hybrids, including notably the polyploid wheats: some have four sets of chromosomes (tetraploid) or six (hexaploid), while other wheat species have (like most eukaryotic organisms) two sets (diploid), so hybridization events likely involved the doubling of chromosome sets, causing immediate genetic isolation.

Hybridization may be important in speciation in some plant groups. However, homoploid hybrid speciation (not increasing the number of sets of chromosomes) may be rare: by 1997, only eight natural examples had been fully described. Experimental studies suggest that hybridization offers a rapid route to speciation, a prediction confirmed by the fact that early generation hybrids and ancient hybrid species have matching genomes, meaning that once hybridization has occurred, the new hybrid genome can remain stable.

Many hybrid zones are known where the ranges of two species meet, and hybrids are continually produced in great numbers. These hybrid zones are useful as biological model systems for studying the mechanisms of speciation. Recently DNA analysis of a bear shot by a hunter in the Northwest Territories confirmed the existence of naturally occurring and fertile grizzly–polar bear hybrids.

Hybrid Vigour

Hybridization between reproductively isolated species often results in hybrid offspring with lower fitness than either parental. However, hybrids are not, as might be expected, always intermediate between their parents (as if there were blending inheritance), but are sometimes stronger or perform better than either parental lineage or variety, a phenomenon called heterosis, hybrid vigour, or heterozygote advantage. This is most common with plant hybrids. A transgressive phenotype is a phenotype that displays more extreme characteristics than either of the parent lines. Plant breeders use several techniques to produce hybrids, including line breeding and the formation of complex hybrids. An economically important example is hybrid maize (corn), which provides a considerable seed yield advantage over open pollinated varieties. Hybrid seed dominates the commercial maize seed market in the United States, Canada and many other major maize-producing countries.

In a hybrid, any trait that falls outside the range of parental variation (and is thus not simply intermediate between its parents) is considered heterotic. *Positive heterosis* produces more robust hybrids, they might be stronger or bigger; while the term *negative heterosis* refers to weaker or smaller hybrids. Heterosis is common in both animal and plant hybrids. For example, hybrids between a lion and a tigress ("ligers") are much larger than either of the two progenitors, while "tigons" (lioness × tiger) are smaller. Similarly, the hybrids between the common pheasant (*Phasianus colchicus*) and domestic fowl (*Gallus gallus*) are larger than either of their parents, as are those produced between the common pheasant and hen golden pheasant (*Chrysolophus pictus*). Spurs are absent in hybrids of the former type, although present in both parents.

4.8.4 Human Influence

Anthropogenic Hybridization

Hybridization is greatly influenced by human impact on the environment, through effects such as habitat fragmentation and species introductions. Such impacts make it difficult to conserve the genetics of populations undergoing introgressive hybridization. Humans have introduced species worldwide to environments for a long time, both intentionally for purposes such as biological control, and unintentionally, as with accidental escapes of individuals. Introductions can drastically affect populations, including through hybridization.

Management

There is a kind of continuum with three semi-distinct categories dealing with anthropogenic hybridization: hybridization without introgression, hybridization with widespread introgression (backcrossing with one of the parent species), and hybrid swarms (highly variable populations with much interbreeding as well as backcrossing with the parent species). Depending on where a population falls along this continuum, the management plans for that population will change. Hybridization is currently an area of great discussion within wildlife management and habitat management. Global climate change is creating other changes such as difference in population distributions which are indirect causes for an increase in anthropogenic hybridization.

Conservationists disagree on when is the proper time to give up on a population that is becoming a hybrid swarm, or to try and save the still existing pure individuals. Once a population becomes a complete mixture, the goal becomes to conserve those hybrids to avoid their loss. Conservationists treat each case on its merits, depending on detecting hybrids within the population. It is nearly impossible to formulate a uniform hybridization policy, because hybridization can occur beneficially when it occurs "naturally", and when hybrid swarms are the only remaining evidence of prior species, they need to be conserved as well.

Genetic Mixing and Extinction

Regionally developed ecotypes can be threatened with extinction when new alleles or genes are introduced that alter that ecotype. This is sometimes called genetic mixing. Hybridization and introgression, which can happen in natural and hybrid populations, of new genetic material can lead to the replacement of local genotypes if the hybrids are more fit and have breeding advantages over the indigenous ecotype or species. These hybridization events can result from the introduction of non-native genotypes by humans or through habitat modification, bringing previously isolated species into contact. Genetic mixing can be especially detrimental for rare species in isolated habitats, ultimately affecting the population to such a degree that none of the originally genetically distinct population remains.

Effect on Biodiversity and Food Security

In agriculture and animal husbandry, the Green Revolution's use of conventional hybridization increased yields by breeding high-yielding varieties. The replacement of locally indigenous breeds, compounded with unintentional cross-pollination and crossbreeding (genetic mixing), has reduced the gene pools of various wild and indigenous breeds resulting in the loss of genetic diversity. Since the indigenous breeds are often well-adapted to local extremes in climate and have immunity to local pathogens, this can be a significant genetic erosion of the gene pool for future breeding. Therefore, commercial plant geneticists strive to breed "widely adapted" cultivars to counteract this tendency.

4.9 CONCLUSION

The tools of recombinant DNA technology have opened up countless possibilities in science and industry, from genetic engineering of crops to the development of gene therapies. Each tool, from restriction enzymes and DNA ligase to PCR and DNA sequencing, plays a specific role in manipulating and studying genetic material, enabling scientists to create novel DNA constructs, express proteins, and study gene function in unprecedented detail. The ongoing development and refinement of these tools continue to expand the frontiers of biotechnology, offering new solutions to medical, agricultural, and environmental challenges.

CHAPTER 5
FUNDAMENTAL FEATURES OF EUKARYOTIC GENES

5.1 INTRODUCTION TO EUKARYOTIC GENES

Eukaryotic genes are the units of heredity in complex organisms, such as plants, animals, fungi, and protists. They encode instructions for synthesizing proteins and other molecules essential for cellular structure and function. Unlike prokaryotic genes, eukaryotic genes are organized and regulated in unique ways to accommodate the complexity of multicellular life. This involves specialized mechanisms for gene expression, regulatory elements, and post-transcriptional modifications.

5.2 GENE STRUCTURE: INTRONS AND EXONS

A defining characteristic of eukaryotic genes is their interrupted structure, consisting of coding sequences (exons) interspersed with non-coding sequences (introns). Exons contain the sequences that code for amino acids, while introns are non-coding segments that are removed during the process of RNA splicing.

This arrangement allows for alternative splicing, whereby different exons can be combined in multiple ways to create diverse protein variants from a single gene. This greatly expands the functional repertoire of eukaryotic genomes and allows organisms to adapt to various physiological and environmental challenges.

The discovery of introns and exons was revolutionary, as it revealed that genes are not continuous sequences of DNA. Instead, eukaryotic genes are modular, with each exon potentially encoding specific functional domains of proteins. The presence of introns and exons also suggests an evolutionary advantage, allowing organisms to evolve new gene functions through exon shuffling or recombination events without disrupting the entire gene sequence.

5.2.1 Intron

An intron is a non-coding sequence in a gene. It is any nucleotide sequence within a gene that is removed by RNA splicing to get the final RNA product of a gene. The term *intron* refers to both the DNA sequence within a gene, and the corresponding sequence in RNA transcripts.

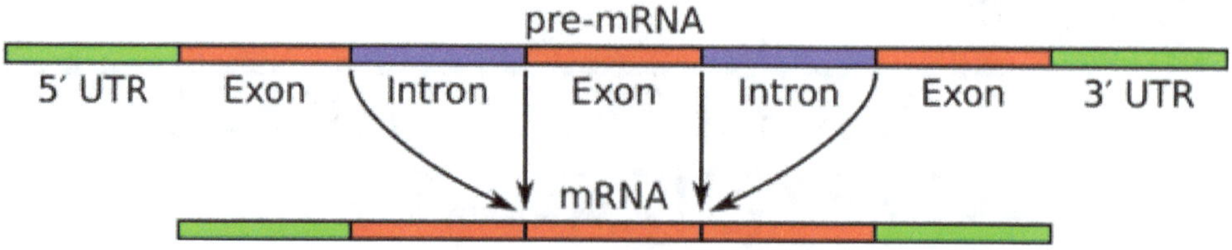

Fig: A Spliceosome Removes Introns from a Transcribed Pre-mRNA Segment (Top). This is Called 'Splicing'. After the Introns have been Removed (Bottom), the Mature mRNA Sequence is Ready for Translation

Sequences of coding DNA which are joined together in the final RNA after RNA splicing are exons. They code for amino acids in the final polypeptide.

Introns are in the genes of most organisms and many viruses. They can be in a wide range of genes, including those that generate proteins, ribosomal RNA (rRNA), and transfer RNA (tRNA). RNA splicing takes place after transcription and before translation.

- **Introns:** Parts of a gene which are discarded: non-working bits.
- **Exons:** Parts of a gene which are expressed: bits of a gene which code for amino-acid sequences in a protein.

The discovery of introns led to the Nobel Prize in Physiology or Medicine in 1993 for Phillip Sharp and Richard Roberts. The term *intron* was introduced by American biochemist Walter Gilbert.

Biological Meaning

There are many unanswered questions about introns. It is unclear whether introns serve some specific function, or whether they are selfish DNA which reproduces itself as a parasite. Recent studies of entire eukaryotic genomes have now shown that the lengths and density (introns/gene) of introns varies considerably between related species. There are four or five different kinds of intron. Some introns represent mobile genetic elements (transposons). Alternative splicing of introns within a gene allows a variety of protein isoforms from a single gene. Thus multiple related proteins can be generated from a single gene and a single precursor mRNA transcript. The control of alternative RNA splicing is performed by complex network of signalling molecules. In humans, ~95% of genes with more than one exon are alternatively spliced.

5.2.2 Exon

An exon is any part of a gene that will form a part of the final mature RNA produced by that gene after introns have been removed by RNA splicing. The term *exon* refers to both the DNA sequence within a gene and to the corresponding sequence in RNA transcripts. In RNA splicing, introns are removed and exons are covalently joined to one another as part of generating the mature RNA. Just as the entire set of genes for a species constitutes the genome, the entire set of exons constitutes the exome.

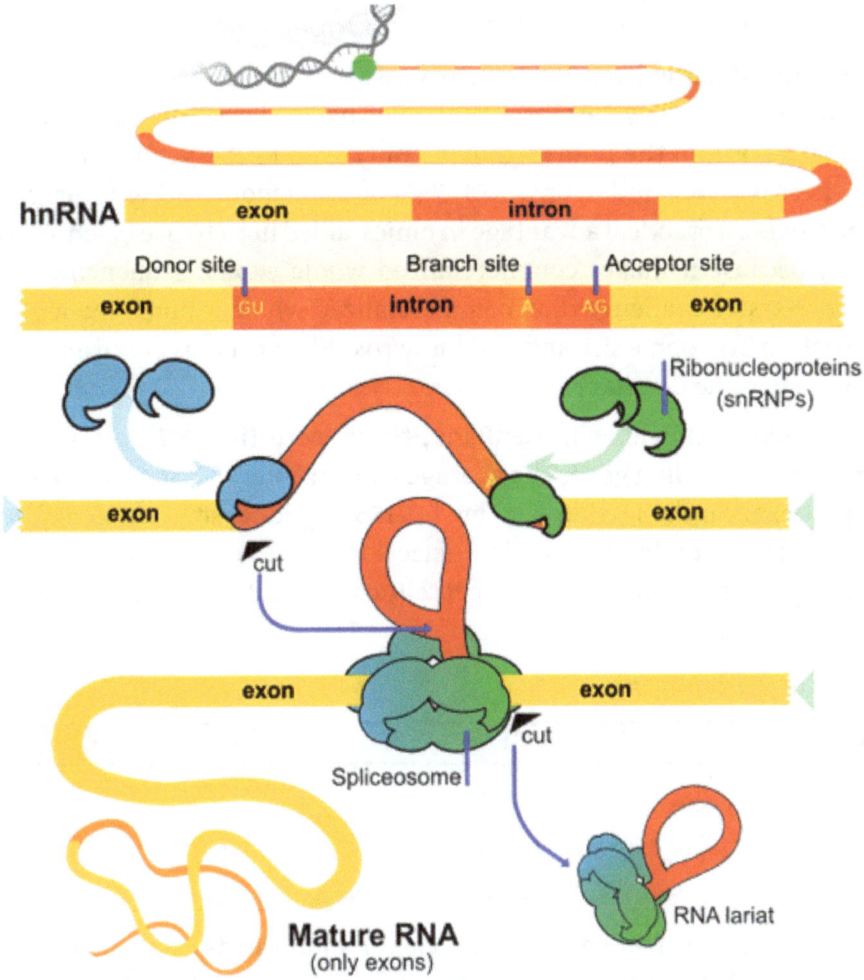

Fig: Introns are Removed and Exons Joined in the Process of RNA Splicing. RNAs could be mRNA or Non-Coding RNA

History

The term *exon* derives from the expressed region and was coined by American biochemist Walter Gilbert in 1978: "The notion of the cistron... must be replaced by that of a transcription unit containing regions which will be lost from the mature messenger – which I suggest we call introns (for intragenic regions) – alternating with regions which will be expressed – exons."

This definition was originally made for protein-coding transcripts that are spliced before being translated. The term later came to include sequences removed from rRNA and tRNA, and other ncRNA and it also was used later for RNA molecules originating from different parts of the genome that are then ligated by trans-splicing.

Contribution to Genomes and Size Distribution

Although unicellular eukaryotes such as yeast have either no introns or very few, metazoans and especially vertebrate genomes have a large fraction of non-coding DNA. For instance, in the human genome only 1.1% of the genome is spanned by exons, whereas 24% is in introns, with 75% of the genome being intergenic DNA. This can provide a practical advantage in omics-aided health care (such as precision medicine) because it makes commercialized whole exome sequencing a smaller and less expensive challenge than commercialized whole genome sequencing. The large variation in genome size and C-value across life forms has posed an interesting challenge called the C-value enigma.

Across all eukaryotic genes in GenBank, there were (in 2002), on average, 5.48 exons per protein coding gene. The average exon encoded 30-36 amino acids. While the longest exon in the human genome is 11555 bp long, several exons have been found to be only 2 bp long. A single-nucleotide exon has been reported from the *Arabidopsis* genome. In humans, like protein coding mRNA, most non-coding RNA also contain multiple exons

Structure and Function

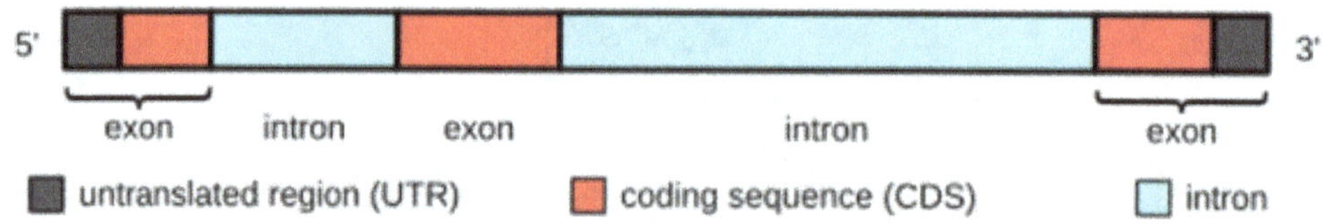

Fig: Exons in a Messenger RNA Precursor (Pre-mRNA).

Exons can include both sequences that code for amino acids (red) and untranslated sequences (grey). Introns — those parts of the pre-mRNA that are not in the mRNA — (blue) are removed, and the exons are joined (spliced) to form the final functional mRNA. The 5′ and 3′ ends of the mRNA are marked to differentiate the two untranslated regions (grey).

In protein-coding genes, the exons include both the protein-coding sequence and the 5′- and 3′-untranslated regions (UTR). Often the first exon includes both the 5′-UTR and the first part of the coding sequence, but exons containing only regions of 5′-UTR or (more rarely) 3′-UTR occur in some genes, i.e., the UTRs may contain introns. Some non-coding RNA transcripts also have exons and introns. Mature mRNAs originating from the same gene need not include the same exons, since different introns in the pre-mRNA can be removed by the process of alternative splicing. Exonization is the creation of a new exon, as a result of mutations in introns.

Experimental Approaches Using Exons

Exon trapping or 'gene trapping' is a molecular biology technique that exploits the existence of the intron-exon splicing to find new genes. The first exon of a 'trapped' gene splices into the exon that is contained in the insertional DNA. This new exon contains the ORF for a reporter gene that can now be expressed using the enhancers that control the target gene. A scientist knows that a new gene has been trapped when the reporter gene is expressed.

Splicing can be experimentally modified so that targeted exons are excluded from mature mRNA transcripts by blocking the access of splice-directing small nuclear ribonucleoprotein particles (snRNPs) to pre-mRNA using Morpholino antisense oligos. This has become a standard technique in developmental biology. Morpholino oligos can also be targeted to prevent molecules that regulate splicing (e.g., splice enhancers, splice suppressors) from binding to pre-mRNA, altering patterns of splicing.

Common Misuse of the Term

Common incorrect uses of the term *exon* are that 'exons code for protein', or 'exons code for amino-acids' or 'exons are translated'. However, these sorts of definitions only cover protein-coding genes, and omit those exons that become part of a non-coding RNA or the untranslated region of an mRNA. Such incorrect definitions still occur in overall reputable secondary sources.

5.3 PROMOTERS: INITIATING TRANSCRIPTION

In genetics, a promoter is a sequence of DNA to which proteins bind to initiate transcription of a single RNA transcript from the DNA downstream of the promoter. The RNA transcript may encode a protein (mRNA), or can have a function in and of itself, such as tRNA or rRNA. Promoters are located near the transcription start sites of genes, upstream on the DNA (towards the 5' region of the sense strand). Promoters can be about 100–1000 base pairs long, the sequence of which is highly dependent on the gene and product of transcription, type or class of RNA polymerase recruited to the site, and species of organism.

Promoters control gene expression in bacteria and eukaryotes. RNA polymerase must attach to DNA near a gene for transcription to occur. Promoter DNA sequences provide an enzyme binding site. The -10 sequence is TATAAT. -35 sequences are conserved on average, but not in most promoters.

Artificial promoters with conserved -10 and -35 elements transcribe more slowly. All DNAs have "Closely spaced promoters". Divergent, tandem, and convergent orientations are possible. Two closely spaced promoters will likely interfere. Regulatory elements can be several kilobases away from the transcriptional start site in gene promoters (enhancers).

In eukaryotes, the transcriptional complex can bend DNA, allowing regulatory sequences to be placed far from the transcription site. The distal promoter is upstream of the gene and may contain additional regulatory elements with a weaker influence. RNA polymerase II (RNAP II) bound to the transcription start site promoter can start mRNA synthesis. It also typically contains CpG islands, a TATA box, and TFIIB recognition elements.

Hypermethylation downregulates both genes, while demethylation upregulates them. Non-coding RNAs are linked to mRNA promoter regions. Subgenomic promoters range from 24 to 100 nucleotides (Beet necrotic yellow vein virus). Gene expression depends on promoter binding. Unwanted gene changes can increase a cell's cancer risk.

MicroRNA promoters often contain CpG islands. DNA methylation forms 5-methylcytosines at the 5' pyrimidine ring of CpG cytosine residues. Some cancer genes are silenced by mutation, but most are silenced by DNA methylation. Others are regulated promoters. Selection may favor less energetic transcriptional binding.

Variations in promoters or transcription factors cause some diseases. Misunderstandings can result from using a canonical sequence to describe a promoter.

For transcription to take place, the enzyme that synthesizes RNA, known as RNA polymerase, must attach to the DNA near a gene. Promoters contain specific DNA sequences such as response elements that provide a secure initial binding site for RNA polymerase and for proteins called transcription factors that recruit RNA polymerase. These transcription factors have specific activator or repressor sequences of corresponding nucleotides that attach to specific promoters and regulate gene expression.

- **In Bacteria:** The promoter is recognized by RNA polymerase and an associated sigma factor, which in turn are often brought to the promoter DNA by an activator protein's binding to its own DNA binding site nearby.
- **In Eukaryotes:** The process is more complicated, and at least seven different factors are necessary for the binding of an RNA polymerase II to the promoter.

Promoters represent critical elements that can work in concert with other regulatory regions (enhancers, silencers, boundary elements/insulators) to direct the level of transcription of a given gene. A promoter is induced in response to changes in abundance or conformation of regulatory proteins in a cell, which enable activating transcription factors to recruit RNA polymerase.

Given the short sequences of most promoter elements, promoters can rapidly evolve from random sequences.

For instance, in *E. coli*, ~60% of random sequences can evolve expression levels comparable to the wild-type lac promoter with only one mutation, and that ~10% of random sequences can serve as active promoters even without evolution.

5.3.1 Identification of Relative Location

As promoters are typically immediately adjacent to the gene in question, positions in the promoter are designated relative to the transcriptional start site, where transcription of DNA begins for a particular gene (i.e., positions upstream are negative numbers counting back from -1, for example -100 is a position 100 base pairs upstream).

5.3.2 Elements

Bacterial

In bacteria, the promoter contains two short sequence elements approximately 10 (Pribnow Box) and 35 nucleotides *upstream* from the transcription start site.

- The sequence at -10 (the -10 element) has the consensus sequence TATAAT.
- The sequence at -35 (the -35 element) has the consensus sequence TTGACA.
- The consensus sequences, while conserved on average, are not found intact in most promoters. On average, only 3 to 4 of the 6 base pairs in each consensus sequence are found in any given promoter. Few natural promoters have been identified to date that possess intact consensus sequences at both the -10 and -35; artificial promoters with complete conservation of the -10 and -35 elements have been found to transcribe at lower frequencies than those with a few mismatches with the consensus.
- The optimal spacing between the -35 and -10 sequences is 17 bp.
- Some promoters contain one or more upstream promoter element (UP element) subsites (consensus sequence 5'-AAAAAARNR-3' when centered in the -42 region; consensus sequence 5'-AWWWWWTTTTT-3' when centered in the -52 region; W = A or T; R = A or G; N = any base).

The promoter sequences are recognized only by RNA polymerase holoenzyme containing sigma-70. RNA polymerase holoenzymes containing other sigma factors recognize different core promoter sequences.

```
← upstream                                                                                      downstream →
5'-XXXXXXXPPPPPPXXXXXXPPPPPPXXXXGGGGGGGGGGGGGGGGGGGGGGGGGGGGGGGGGGGGGGGGGGGGGGGGGGGGGGGGGGGGGGGGXXXX-3'
         -35         -10          Gene to be transcribed
```

Probability of Occurrence of Each Nucleotide

```
for -10 sequence
 T    A    T    A    A    T
77%  76%  60%  61%  56%  82%

for -35 sequence
 T    T    G    A    C    A
69%  79%  61%  56%  54%  54%
```

Bidirectional (Prokaryotic)

Promoters can be very closely located in the DNA. Such “closely spaced promoters” have been observed in the DNAs of all life forms, from humans to prokaryotes and are highly conserved. Therefore, they may provide some (presently unknown) advantages. These pairs of promoters can be positioned in divergent, tandem, and convergent directions.

They can also be regulated by transcription factors and differ in various features, such as the nucleotide distance between them, the two promoter strengths, etc. The most important aspect of two closely spaced promoters is that they will, most likely, interfere with each other. Several studies have explored this using both analytical and stochastic models. There are also studies that measured gene expression in synthetic genes or from one to a few genes controlled by bidirectional promoters.

More recently, one study measured most genes controlled by tandem promoters in *E. coli*. In that study, two main forms of interference were measured. One is when an RNAP is on the downstream promoter, blocking the movement of RNAPs elongating from the upstream promoter.

The other is when the two promoters are so close that when an RNAP sits on one of the promoters, it blocks any other RNAP from reaching the other promoter. These events are possible because the RNAP occupies several nucleotides when bound to the DNA, including in transcription start sites. Similar events occur when the promoters are in divergent and convergent formations. The possible events also depend on the distance between them.

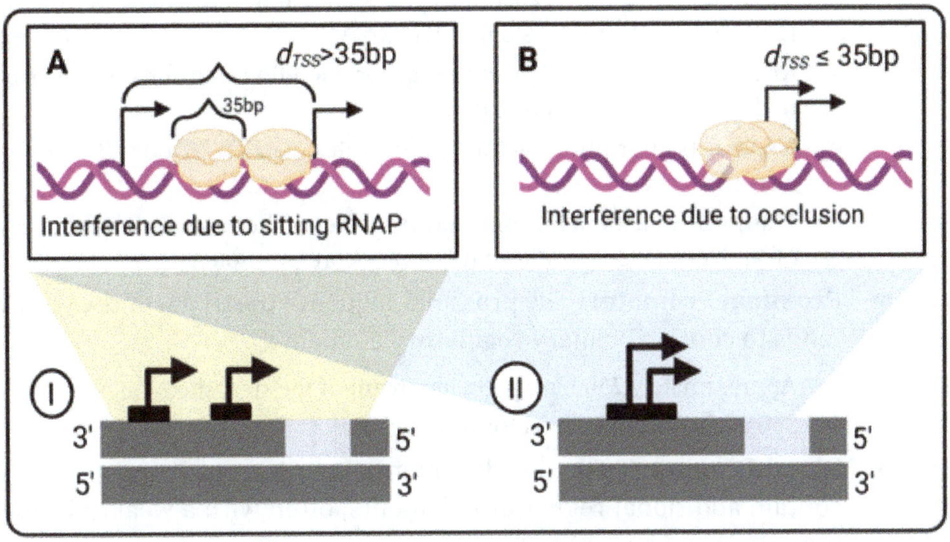

Fig: Depiction the Phenomenon of Interference between Tandem Promoters

Eukaryotic

Gene promoters are typically located upstream of the gene and can have regulatory elements several kilobases away from the transcriptional start site (enhancers). In eukaryotes, the transcriptional complex can cause the DNA to bend back on itself, which allows for placement of regulatory sequences far from the actual site of transcription. Eukaryotic RNA-polymerase-II-dependent promoters can contain a TATA box (consensus sequence TATAAA), which is recognized by the general transcription factor TATA-binding protein (TBP); and a B recognition element (BRE), which is recognized by the general transcription factor TFIIB. The TATA element and BRE typically are located close to the transcriptional start site (typically within 30 to 40 base pairs).

Eukaryotic promoter regulatory sequences typically bind proteins called transcription factors that are involved in the formation of the transcriptional complex. An example is the E-box (sequence CACGTG), which binds transcription factors in the basic helix-loop-helix (bHLH) family (e.g. BMAL1-Clock, cMyc). Some promoters that are targeted by multiple transcription factors might achieve a hyperactive state, leading to increased transcriptional activity.

- **Core Promoter:** The minimal portion of the promoter required to properly initiate transcription.
  - Includes the transcription start site (TSS) and elements directly upstream.
  - A binding site for RNA polymerase.

  - *RNA Polymerase I:* Transcribes genes encoding 18S, 5.8S and 28S ribosomal RNAs.
  - *RNA Polymerase II:* Transcribes genes encoding messenger RNA and certain small nuclear RNAs and microRNA.
  - *RNA Polymerase III:* Transcribes genes encoding transfer RNA, 5s ribosomal RNAs and other small RNAs.
  - General transcription factor binding sites, e.g. TATA box, B recognition element.
  - Many other elements/motifs may be present. There is no such thing as a set of "universal elements" found in every core promoter.
- **Proximal Promoter:** The proximal sequence upstream of the gene that tends to contain primary regulatory elements.
  - Approximately 250 base pairs upstream of the start site.
  - Specific transcription factor binding sites.
- **Distal Promoter:** The distal sequence upstream of the gene that may contain additional regulatory elements, often with a weaker influence than the proximal promoter.
  - Anything further upstream (but not an enhancer or other regulatory region whose influence is positional/orientation independent).
  - Specific transcription factor binding sites.

Bidirectional (Mammalian)

Bidirectional promoters are short (<1 kbp) intergenic regions of DNA between the 5' ends of the genes in a bidirectional gene pair. A "bidirectional gene pair" refers to two adjacent genes coded on opposite strands, with their 5' ends oriented toward one another. The two genes are often functionally related, and modification of their shared promoter region allows them to be co-regulated and thus co-expressed. Bidirectional promoters are a common feature of mammalian genomes. About 11% of human genes are bidirectionally paired.

Bidirectionally paired genes in the Gene Ontology database shared at least one database-assigned functional category with their partners 47% of the time. Microarray analysis has shown bidirectionally paired genes to be co-expressed to a higher degree than random genes or neighboring unidirectional genes. Although co-expression does not necessarily indicate co-regulation, methylation of bidirectional promoter regions has been shown to downregulate both genes, and demethylation to upregulate both genes. There are exceptions to this, however. In some cases (about 11%), only one gene of a bidirectional pair is expressed. In these cases, the promoter is implicated in suppression of the non-expressed gene. The mechanism behind this could be competition for the same polymerases, or chromatin modification. Divergent transcription could shift nucleosomes to upregulate transcription of one gene, or remove bound transcription factors to downregulate transcription of one gene.

Some functional classes of genes are more likely to be bidirectionally paired than others. Genes implicated in DNA repair are five times more likely to be regulated by bidirectional promoters than by unidirectional promoters. Chaperone proteins are three times more likely, and mitochondrial genes are more than twice as likely. Many basic housekeeping and cellular metabolic genes are regulated by bidirectional promoters. The overrepresentation of bidirectionally paired DNA repair genes associates these promoters with cancer. Forty-five percent of human somatic oncogenes seem to be regulated by bidirectional promoters - significantly more than non-cancer causing genes. Hypermethylation of the promoters between gene pairs WNT9A/CD558500, CTDSPL/BC040563, and KCNK15/BF195580 has been associated with tumors.

Certain sequence characteristics have been observed in bidirectional promoters, including a lack of TATA boxes, an abundance of CpG islands, and a symmetry around the midpoint of dominant Cs and As on one side and Gs and Ts on the other. A motif with the consensus sequence of TCTCGCGAGA, also called the CGCG element, was recently shown to drive PolII-driven bidirectional transcription in CpG islands. CCAAT boxes are common, as they are in many promoters that lack TATA boxes. In addition, the motifs NRF-1, GABPA, YY1, and ACTACAnnTCCC are represented in bidirectional promoters at significantly higher rates than in unidirectional promoters. The absence of TATA boxes in bidirectional promoters suggests that TATA boxes play a role in determining the directionality of promoters, but counterexamples of bidirectional promoters do possess TATA boxes and unidirectional promoters without them indicates that they cannot be the only factor.

Although the term "bidirectional promoter" refers specifically to promoter regions of mRNA-encoding genes, luciferase assays have shown that over half of human genes do not have a strong directional bias. Research suggests that non-coding RNAs are frequently associated with the promoter regions of mRNA-encoding genes. It has been hypothesized that the recruitment and initiation of RNA polymerase II usually begins bidirectionally, but divergent transcription is halted at a checkpoint later during elongation. Possible mechanisms behind this regulation include sequences in the promoter region, chromatin modification, and the spatial orientation of the DNA.

5.3.3 Subgenomic

A subgenomic promoter is a promoter added to a virus for a specific heterologous gene, resulting in the formation of mRNA for that gene alone. Many positive-sense

RNA viruses produce these subgenomic mRNAs (sgRNA) as one of the common infection techniques used by these viruses and generally transcribe late viral genes. Subgenomic promoters range from 24 nucleotide (Sindbis virus) to over 100 nucleotides (Beet necrotic yellow vein virus) and are usually found upstream of the transcription start.

5.3.4 Detection

A wide variety of algorithms have been developed to facilitate detection of promoters in genomic sequence, and promoter prediction is a common element of many gene prediction methods. A promoter region is located before the -35 and -10 Consensus sequences. The closer the promoter region is to the consensus sequences the more often transcription of that gene will take place. There is not a set pattern for promoter regions as there are for consensus sequences.

5.3.5 Binding

The initiation of the transcription is a multistep sequential process that involves several mechanisms: promoter location, initial reversible binding of RNA polymerase, conformational changes in RNA polymerase, conformational changes in DNA, binding of nucleoside triphosphate (NTP) to the functional RNA polymerase-promoter complex, and nonproductive and productive initiation of RNA synthesis. The promoter binding process is crucial in the understanding of the process of gene expression. Tuning synthetic genetic systems relies on precisely engineered synthetic promoters with known levels of transcription rates.

Location

Although RNA polymerase holoenzyme shows high affinity to non-specific sites of the DNA, this characteristic does not allow us to clarify the process of promoter location. This process of promoter location has been attributed to the structure of the holoenzyme to DNA and sigma 4 to DNA complexes.

5.3.6 Diseases Associated with Aberrant Function

Most diseases are heterogeneous in cause, meaning that one "disease" is often many different diseases at the molecular level, though symptoms exhibited and response to treatment may be identical. How diseases of different molecular origin respond to treatments is partially addressed in the discipline of pharmacogenomics.

Not listed here are the many kinds of cancers involving aberrant transcriptional regulation owing to creation of chimeric genes through pathological chromosomal translocation. Importantly, intervention in the number or structure of promoter-bound proteins is one key to treating a disease without affecting expression of unrelated genes sharing elements with the target gene. Some genes whose change is not desirable are capable of influencing the potential of a cell to become cancerous.

5.3.7 CpG Islands in Promoters

In humans, about 70% of promoters located near the transcription start site of a gene (proximal promoters) contain a CpG island. CpG islands are generally 200 to 2000 base pairs long, have a C:G base pair content >50%, and have regions of DNA where a cytosine nucleotide is followed by a guanine nucleotide and this occurs frequently in the linear sequence of bases along its 5' → 3' direction. Distal promoters also frequently contain CpG islands, such as the promoter of the DNA repair gene *ERCC1*, where the CpG island-containing promoter is located about 5,400 nucleotides upstream of the coding region of the *ERCC1* gene. CpG islands also occur frequently in promoters for functional noncoding RNAs such as microRNAs.

5.3.8 Methylation of CpG Islands Stably Silences Genes

In humans, DNA methylation occurs at the 5' position of the pyrimidine ring of the cytosine residues within CpG sites to form 5-methylcytosines. The presence of multiple methylated CpG sites in CpG islands of promoters causes stable silencing of genes. Silencing of a gene may be initiated by other mechanisms, but this is often followed by methylation of CpG sites in the promoter CpG island to cause the stable silencing of the gene.

5.3.9 Promoter CpG Hyper/Hypo-Methylation in Cancer

Generally, in progression to cancer, hundreds of genes are silenced or activated. Although silencing of some genes in cancers occurs by mutation, a large proportion of carcinogenic gene silencing is a result of altered DNA methylation. DNA methylation causing silencing in cancer typically occurs at multiple CpG sites in the CpG islands that are present in the promoters of protein coding genes. Altered expressions of microRNAs also silence or activate many genes in progression to cancer. Altered microRNA expression occurs through hyper/hypo-methylation of CpG sites in CpG islands in promoters controlling transcription of the microRNAs. Silencing of DNA repair genes through methylation of CpG islands in their promoters appears to be especially important in progression to cancer.

5.3.10 Canonical Sequences and Wild-Type

The usage of the term canonical sequence to refer to a promoter is often problematic, and can lead to misunderstandings about promoter sequences. Canonical implies perfect, in some sense. In the case of a transcription factor binding site, there may be a single sequence that binds the protein most strongly under specified cellular conditions. This might be called canonical. However, natural selection may favor less energetic binding as a way of regulating transcriptional output. In this case, we may call the most common sequence in a population the wild-type sequence. It may not even be the most advantageous sequence to have under prevailing conditions.

Recent evidence also indicates that several genes (including the proto-oncogene c-myc) have G-quadruplex motifs as potential regulatory signals.

5.3.11 Synthetic Promoter Design and Engineering

Promoters are important gene regulatory elements used in tuning synthetically designed genetic circuits and metabolic networks. For example, to overexpress an important gene in a network, to yield higher production of target protein, synthetic biologists design promoters to upregulate its expression. Automated algorithms can be used to design neutral DNA or insulators that do not trigger gene expression of downstream sequences.

5.4 ENHANCERS AND SILENCERS: REGULATORY ELEMENTS FOR GENE EXPRESSION

Enhancers and silencers are regulatory DNA elements that modulate gene expression from a distance. Enhancers increase the transcription of a gene, often by binding activator proteins, while silencers decrease transcription by binding repressor proteins. These elements can be located far away from the gene they regulate, sometimes even several kilobases downstream or upstream. Enhancers and silencers can loop around to interact with the gene's promoter region, bringing regulatory proteins into proximity with the transcription machinery.

The role of enhancers and silencers in gene expression is crucial for developing complex multicellular organisms, as they allow for finely tuned control of gene activity in response to developmental cues or environmental factors. Enhancers, for example, allow specific genes to be activated in particular tissues or at certain developmental stages, contributing to cellular diversity within an organism. The flexibility in regulatory control afforded by enhancers and silencers enables organisms to adapt gene expression dynamically.

5.4.1 Enhancer

In genetics, an enhancer is a short (50–1500 bp) region of DNA that can be bound by proteins (activators) to increase the likelihood that transcription of a particular gene will occur. These proteins are usually referred to as transcription factors. Enhancers are *cis*-acting. They can be located up to 1 Mbp (1,000,000 bp) away from the gene, upstream or downstream from the start site. There are hundreds of thousands of enhancers in the human genome. They are found in both prokaryotes and eukaryotes. Active enhancers typically get transcribed as enhancer or regulatory non-coding RNA, whose expression levels correlate with mRNA levels of target genes.

The first discovery of a eukaryotic enhancer was in the immunoglobulin heavy chain gene in 1983. This enhancer, located in the large intron, provided an explanation for the transcriptional activation of rearranged Vh gene promoters while unrearranged Vh promoters remained inactive. Lately, enhancers have been shown to be involved in certain medical conditions, for example, myelosuppression. Since 2022, scientists have used artificial intelligence to design synthetic enhancers and applied them in animal systems, first in a cell line, and one year later also in vivo.

Locations

In eukaryotic cells the structure of the chromatin complex of DNA is folded in a way that functionally mimics the supercoiled state characteristic of prokaryotic DNA, so although the enhancer DNA may be far from the gene in a linear way, it is spatially close to the promoter and gene. This allows it to interact with the general transcription factors and RNA polymerase II. The same mechanism holds true for silencers in the eukaryotic genome. Silencers are antagonists of enhancers that, when bound to its proper transcription factors called repressors, repress the transcription of the gene. Silencers and enhancers may be in close proximity to each other or may even be in the same region only differentiated by the transcription factor the region binds to.

An enhancer may be located upstream or downstream of the gene it regulates. Furthermore, an enhancer does not need to be located near the transcription initiation site to affect transcription, as some have been found located several hundred thousand base pairs upstream or downstream of the start site. Enhancers do not act on the promoter region itself, but are bound by activator proteins as first shown by in vivo competition experiments. Subsequently, molecular studies showed direct interactions with transcription factors and cofactors, including the mediator complex, which recruits polymerase II and the general transcription factors which then begin transcribing the genes.

Enhancers can also be found within introns. An enhancer's orientation may even be reversed without affecting its function; additionally, an enhancer may be excised and inserted elsewhere in the chromosome, and still affect gene transcription. That is one reason that introns polymorphisms may have effects although they are not translated. Enhancers can also be found at the exonic region of an unrelated gene and they may act on genes on another chromosome. Enhancers are bound by p300-CBP and their location can be predicted by ChIP-seq against this family of coactivators.

Theories

As of 2005, there are two different theories on the information processing that occurs on enhancers:

- **Enhanceosomes:** Rely on highly cooperative, coordinated action and can be disabled by single point mutations that move or remove the binding sites of individual proteins.

- **Flexible Billboards:** Less integrative, multiple proteins independently regulate gene expression and their sum is read in by the basal transcriptional machinery.

Examples in the Human Genome

HACNS1

HACNS1 (also known as CENTG2 and located in the Human Accelerated Region 2) is a gene enhancer "that may have contributed to the evolution of the uniquely opposable human thumb, and possibly also modifications in the ankle or foot that allow humans to walk on two legs". Evidence to date shows that of the 110,000 gene enhancer sequences identified in the human genome, HACNS1 has undergone the most change during the evolution of humans following the split with the ancestors of chimpanzees.

GADD45G

An enhancer near the gene GADD45g has been described that may regulate brain growth in chimpanzees and other mammals, but not in humans. The GADD45G regulator in mice and chimps is active in regions of the brain where cells that form the cortex, ventral forebrain, and thalamus are located and may suppress further neurogenesis. Loss of the GADD45G enhancer in humans may contribute to an increase of certain neuronal populations and to forebrain expansion in humans.

In Developmental Biology

The development, differentiation and growth of cells and tissues require precisely regulated patterns of gene expression. Enhancers work as cis-regulatory elements to mediate both spatial and temporal control of development by turning on transcription in specific cells and/or repressing it in other cells. Thus, the particular combination of transcription factors and other DNA-binding proteins in a developing tissue controls which genes will be expressed in that tissue. Enhancers allow the same gene to be used in diverse processes in space and time.

Identification and Characterization

Traditionally, enhancers were identified by enhancer trap techniques using a reporter gene or by comparative sequence analysis and computational genomics. In genetically tractable models such as the fruit fly *Drosophila melanogaster*, for example, a reporter construct such as the *lacZ* gene can be randomly integrated into the genome using a P element transposon. If the reporter gene integrates near an enhancer, its expression will reflect the expression pattern driven by that enhancer. Thus, staining the flies for LacZ expression or activity and cloning the sequence surrounding the integration site allows the identification of the enhancer sequence.

The development of genomic and epigenomic technologies, however, has dramatically changed the outlook for cis-regulatory modules (CRM) discovery. Next-generation sequencing (NGS) methods now enable high-throughput functional CRM discovery assays, and the vastly increasing amounts of available data, including large-scale libraries of transcription factor-binding site (TFBS) motifs, collections of annotated, validated CRMs, and extensive epigenetic data across many cell types, are making accurate computational CRM discovery an attainable goal. An example of NGS-based approach called DNase-seq have enabled identification of nucleosome-depleted, or open chromatin regions, which can contain CRM. More recently techniques such as ATAC-seq have been developed which require less starting material.

Nucelosome depleted regions can be identified in vivo through expression of Dam methylase, allowing for greater control of cell-type specific enhancer identification. Computational methods include comparative genomics, clustering of known or predicted TF-binding sites, and supervised machine-learning approaches trained on known CRMs. All of these methods have proven effective for CRM discovery, but each has its own considerations and limitations, and each is subject to a greater or lesser number of false-positive identifications. In the comparative genomics approach, sequence conservation of non-coding regions can be indicative of enhancers. Sequences from multiple species are aligned, and conserved regions are identified computationally.

Identified sequences can then be attached to a reporter gene such as green fluorescent protein or lacZ to determine the *in vivo* pattern of gene expression produced by the enhancer when injected into an embryo. mRNA expression of the reporter can be visualized by *in situ* hybridization, which provides a more direct measure of enhancer activity, since it is not subjected to the complexities of translation and protein folding.

Although much evidence has pointed to sequence conservation for critical developmental enhancers, other work has shown that the function of enhancers can be conserved with little or no primary sequence conservation. For example, the *RET* enhancers in humans have very little sequence conservation to those in zebrafish, yet both species' sequences produce nearly identical patterns of reporter gene expression in zebrafish.

Similarly, in highly diverged insects (separated by around 350 million years), similar gene expression patterns of several key genes was found to be regulated through similarly constituted CRMs although these CRMs do not show any appreciable sequence conservation detectable by standard sequence alignment methods such as BLAST.

In Segmentation of Insects

The enhancers determining early segmentation in *Drosophila melanogaster* embryos are among the best characterized developmental enhancers. In the early fly embryo, the gap gene transcription factors are responsible for activating and repressing a number of segmentation genes, such as the pair rule genes. The gap genes are expressed in blocks along the anterior-posterior axis of the fly along with other maternal effect transcription factors, thus creating zones within which different combinations of transcription factors are expressed. The pair-rule genes are separated from one another by non-expressing cells. Moreover, the stripes of expression for different pair-rule genes are offset by a few cell diameters from one another. Thus, unique combinations of pair-rule gene expression create spatial domains along the anterior-posterior axis to set up each of the 14 individual segments. The 480 bp enhancer responsible for driving the sharp stripe two of the pair-rule gene *even-skipped* (*eve*) has been well-characterized. The enhancer contains 12 different binding sites for maternal and gap gene transcription factors. Activating and repressing sites overlap in sequence. *Eve* is only expressed in a narrow stripe of cells that contain high concentrations of the activators and low concentration of the repressors for this enhancer sequence. Other enhancer regions drive *eve* expression in 6 other stripes in the embryo.

In Vertebrate Patterning

Establishing body axes is a critical step in animal development. During mouse embryonic development, Nodal, a transforming growth factor-beta superfamily ligand, is a key gene involved in patterning both the anterior-posterior axis and the left-right axis of the early embryo. The *Nodal* gene contains two enhancers: the Proximal Epiblast Enhancer (PEE) and the Asymmetric Enhancer (ASE). The PEE is upstream of the Nodal gene and drives *Nodal* expression in the portion of the primitive streak that will differentiate into the node (also referred to as the primitive node). The PEE turns on Nodal expression in response to a combination of Wnt signaling plus a second, unknown signal; thus, a member of the LEF/TCF transcription factor family likely binds to a TCF binding site in the cells in the node. Diffusion of Nodal away from the node forms a gradient which then patterns the extending anterior-posterior axis of the embryo. The ASE is an intronic enhancer bound by the fork head domain transcription factor Fox1. Early in development, Fox1-driven Nodal expression establishes the visceral endoderm. Later in development, Fox1 binding to the ASE drives *Nodal* expression on the left side of the lateral plate mesoderm, thus establishing left-right asymmetry necessary for asymmetric organ development in the mesoderm.

Establishing three germ layers during gastrulation is another critical step in animal development. Each of the three germ layers has unique patterns of gene expression that promote their differentiation and development. The endoderm is specified early in development by *Gata4* expression, and Gata4 goes on to direct gut morphogenesis later. *Gata4* expression is controlled in the early embryo by an intronic enhancer that binds another forkhead domain transcription factor, FoxA2. Initially the enhancer drives broad gene expression throughout the embryo, but the expression quickly becomes restricted to the endoderm, suggesting that other repressors may be involved in its restriction. Late in development, the same enhancer restricts expression to the tissues that will become the stomach and pancreas. An additional enhancer is responsible for maintaining *Gata4* expression in the endoderm during the intermediate stages of gut development.

Multiple Enhancers Promote Developmental Robustness

Some genes involved in critical developmental processes contain multiple enhancers of overlapping function. Secondary enhancers, or "shadow enhancers", may be found many kilobases away from the primary enhancer ("primary" usually refers to the first enhancer discovered, which is often closer to the gene it regulates). On its own, each enhancer drives nearly identical patterns of gene expression. Are the two enhancers truly redundant? Recent work has shown that multiple enhancers allow fruit flies to survive environmental perturbations, such as an increase in temperature. When raised at an elevated temperature, a single enhancer sometimes fails to drive the complete pattern of expression, whereas the presence of both enhancers permits normal gene expression.

Evolution of Developmental Mechanisms

One theme of research in evolutionary developmental biology ("evo-devo") is investigating the role of enhancers and other cis-regulatory elements in producing morphological changes via developmental differences between species.

Stickleback Pitx1

Recent work has investigated the role of enhancers in morphological changes in threespine stickleback fish. Sticklebacks exist in both marine and freshwater environments, but sticklebacks in many freshwater populations have completely lost their pelvic fins (appendages homologous to the posterior limb of tetrapods). *Pitx1* is a homeobox gene involved in posterior limb development in vertebrates. Preliminary genetic analyses indicated that changes in the expression of this gene were responsible for pelvic reduction in sticklebacks. Fish expressing only the freshwater allele of *Pitx1* do not have pelvic spines, whereas fish expressing a marine allele retain pelvic spines. A more thorough characterization showed that a 500 base pair enhancer sequence is responsible for turning on *Pitx1* expression in the posterior fin bud. This enhancer is located near a chromosomal fragile site—a sequence of DNA that is likely to be broken and thus more likely to be mutated as a result of imprecise DNA repair. This fragile site has caused repeated, independent losses of the enhancer responsible for driving *Pitx1* expression in the pelvic spines in isolated freshwater population, and without this enhancer, freshwater fish fail to develop pelvic spines.

In Drosophila Wing Pattern Evolution

Pigmentation patterns provide one of the most striking and easily scored differences between different species of animals. Pigmentation of the *Drosophila* wing has proven to be a particularly amenable system for studying the development of complex pigmentation phenotypes. The *Drosophila guttifera* wing has 12 dark pigmentation spots and 4 lighter gray intervein patches. Pigment spots arise from expression of the *yellow* gene, whose product produces black melanin. Recent work has shown that two enhancers in the *yellow* gene produce gene expression in precisely this pattern - the vein spot enhancer drives reporter gene expression in the 12 spots, and the intervein shade enhancer drives reporter expression in the 4 distinct patches. These two enhancers are responsive to the Wnt signaling pathway, which is activated by *wingless* expression at all of the pigmented locations. Thus, in the evolution of the complex pigmentation phenotype, the *yellow* pigment gene evolved enhancers responsive to the wingless signal and *wingless* expression evolved at new locations to produce novel wing patterns.

In Inflammation and Cancer

Each cell typically contains several hundred of a special class of enhancers that stretch over many kilobases long DNA sequences, called “super-enhancers”. These enhancers contain a large number of binding sites for sequence-specific, inducible transcription factors, and regulate expression of genes involved in cell differentiation. During inflammation, the transcription factor NF-κB facilitates remodeling of chromatin in a manner that selectively redistributes cofactors from high-occupancy enhancers, thereby repressing genes involved in maintaining cellular identify whose expression they enhance; at the same time, this F-κB-driven remodeling and redistribution activates other enhancers that guide changes in cellular function through inflammation. As a result, inflammation reprograms cells, altering their interactions with the rest of tissue and with the immune system. In cancer, proteins that control NF-κB activity are dysregulated, permitting malignant cells to decrease their dependence on interactions with local tissue, and hindering their surveillance by the immune system.

Designing Enhancers in Synthetic Biology

Synthetic regulatory elements such as enhancers promise to be a powerful tool to direct gene products to particular cell types in order to treat disease by activating beneficial genes or by halting aberrant cell states. Since 2022, artificial intelligence and transfer learning strategies have led to a better understanding of the features of regulatory DNA sequences, the prediction, and the design of synthetic enhancers. Building on work in cell culture, synthetic enhancers were successfully applied to entire living organisms in 2023. Using deep neural networks, scientists simulated the evolution of DNA sequences to analyze the emergence of features that underly enhancer function. This allowed the design and production of a range of functioning synthetic enhancers for different cell types of the fruit fly brain. A second approach trained artificial intelligence models on single-cell DNA accessibility data and transferred the learned models towards the prediction of enhancers for selected tissues in the fruit fly embryo. These enhancer prediction models were used to design synthetic enhancers for the nervous system, brain, muscle, epidermis and gut.

5.4.2 Silencer

In genetics, a silencer is a DNA sequence capable of binding transcription regulation factors, called repressors. DNA contains genes and provides the template to produce messenger RNA (mRNA). That mRNA is then translated into proteins. When a repressor protein binds to the silencer region of DNA, RNA polymerase is prevented from transcribing the DNA sequence into RNA. With transcription blocked, the translation of RNA into proteins is impossible. Thus, silencers prevent genes from being expressed as proteins.

RNA polymerase, a DNA-dependent enzyme, transcribes the DNA sequences, called nucleotides, in the 3' to 5' direction while the complementary RNA is synthesized in the 5' to 3' direction. RNA is similar to DNA, except that RNA contains uracil, instead of thymine, which forms a base pair with adenine. An important region for the activity of gene repression and expression found in RNA is the 3' untranslated region. This is a region on the 3' terminus of RNA that will not be translated to protein but includes many regulatory regions. Not much is yet known about silencers but scientists continue to study in hopes to classify more types, locations in the genome, and diseases associated with silencers.

Functionality

Locations within the Genome

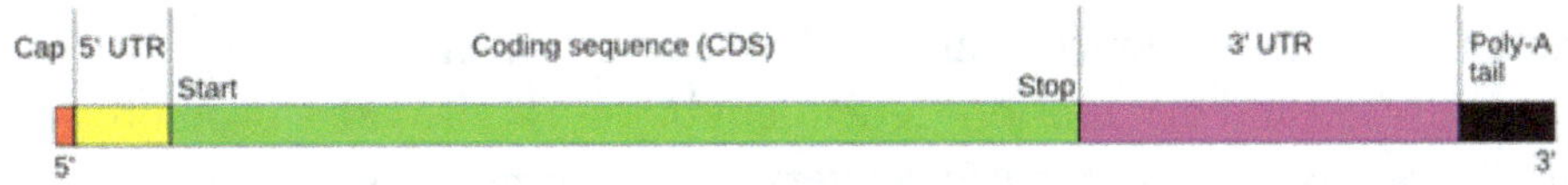

Fig: The 3' Untranslated Region of mRNA Labeled 3' UTR. Normally about 700 Nucleotides in Human mRNA

A silencer is a sequence-specific element that induces a negative effect on the transcription of its particular gene. There are many positions in which a silencer element can be located in DNA.

The most common position is found upstream of the target gene where it can help repress the transcription of the gene. This distance can vary greatly between approximately -20 bp to -2000 bp upstream of a gene. Certain silencers can be found downstream of a promoter located within the intron or exon of the gene itself. Silencers have also been found within the 3 prime untranslated region (3' UTR) of mRNA.

Types

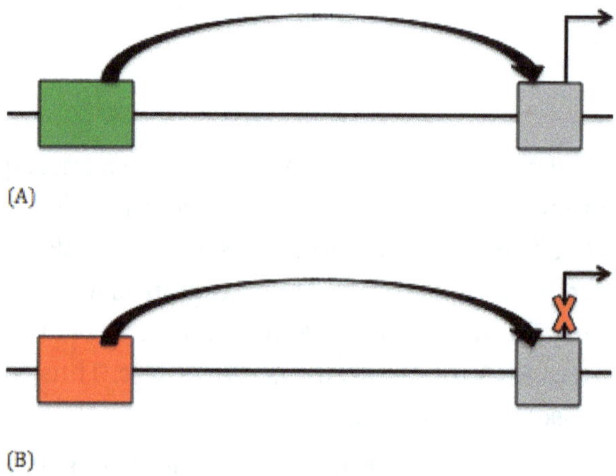

Fig: A Simple Image of how an Enhancer and a Silencer Affect the Function of a Promoter Region

Currently, there are two main types of silencers in DNA, which are the classical silencer element and the non-classical negative regulatory element (NRE). In classical silencers, the gene is actively repressed by the silencer element, mostly by interfering with general transcription factor (GTF) assembly. NREs passively repress the gene, usually by inhibiting other elements that are upstream of the gene. Of the NREs, there are certain silencers that are orientation-dependent meaning that the binding factor binds in a particular direction relative to other sequences.

Promoter-dependent silencers are understood to be silencer elements because they are position and orientation-dependent but must also use a promoter-specific factor. There has been a recent discovery of Polycomb-group Response Elements (PREs), which can allow and inhibit repression depending on the protein bound to it, and the presence of non-coding transcription.

Mechanisms

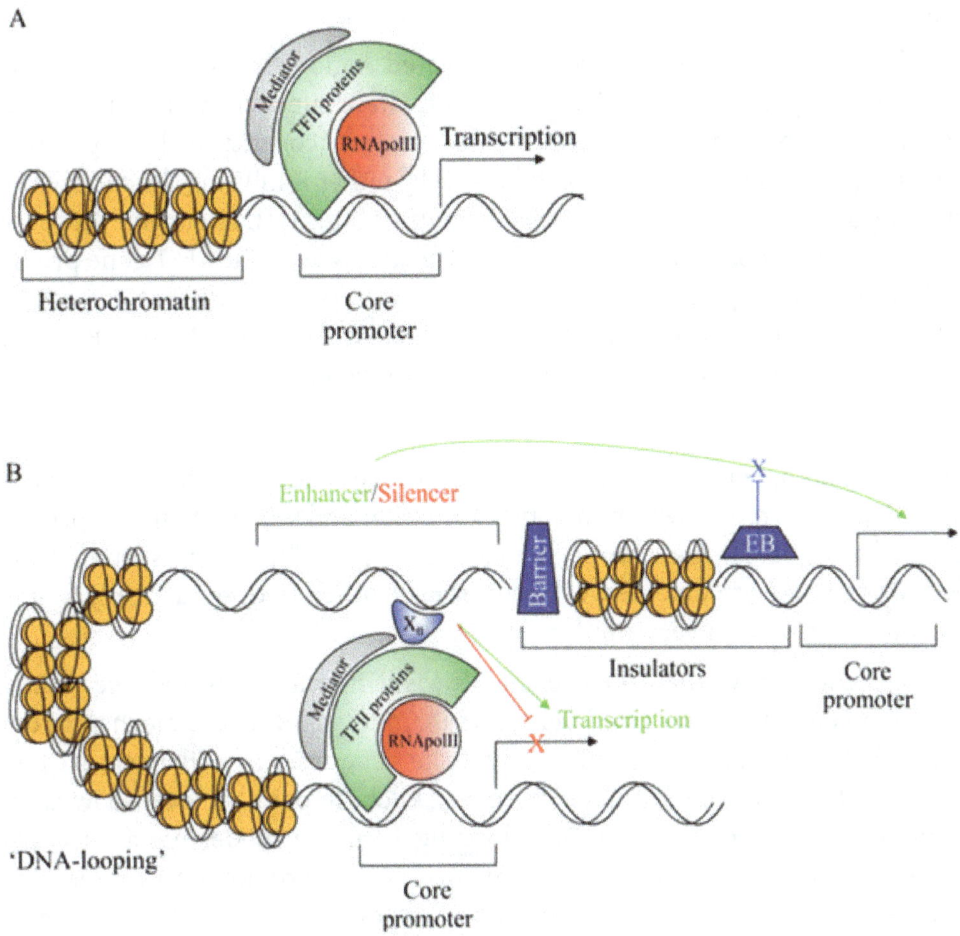

Fig: DNA Silencer Sequence Region and Eukaryotic Transcriptional Machinery

For classical silencers, the signaling pathway is relatively simple. Since repression is active, silencer elements target the assembly of GTFs, necessary for transcription of the gene. These silencer elements are mostly located upstream of the gene and can vary between short and long distances. For long-range silencers, it has been observed that the DNA will form a loop in order to bring the silencer closer to the promoter and loop out the interfering DNA. Silencers also target helicase sites in the DNA that are rich in adenine and thymine (AT) and prone to unwinding the DNA, allowing room to initiate transcription.

The inhibited helicase activity leads to the inhibition of transcription. This is commonly seen in the human thyrotropin-β gene promoter. NREs can induce a bend in the promoter region to block interactions and flank regulatory signals or promoter regions as well. When the silencer region is located within an intron, there can be two types of repressions. First, there can be a physical blockage of a splice site. Second, there can be a bend in the DNA that will inhibit RNA processing.

When located in the exon or the untranslated region, the silencer will mainly be classical or position-dependent. However, these silencers can carry out their activity prior to transcription. Most silencers are constitutively expressed in organisms, only allowing activation of a gene by either inhibiting the silencer or by activating an enhancer region. The best example of this is the Neuronal-Restrictive Silencer Factor (NRSF) that is produced by the *REST* gene. The *REST* gene produces NRSF in order to repress the transcription of neuronal genes that are essential for localization of neuronal tissue. When a silencer represses *REST*, NRSF is also inhibited, allowing for the transcription of neuronal genes.

Similarities with Enhancers

Another regulatory element located upstream of the gene is an enhancer. Enhancers function as a "turn on" switch in gene expression and will activate the promoter region of a particular gene while silencers act as the "turn off" switch. Though these two regulatory elements work against each other, both sequence types affect the promoter region in very similar ways.

Because silencers have not been thoroughly identified and analyzed, the extensive research on enhancers has aided biologists in understanding the mechanics of the silencer. Enhancers can be found in many of the same areas that silencers are found, such as upstream of the promoter by many kilobase pairs, or even downstream within the intron of the gene. DNA looping is also a model function used by enhancers in order to shorten the proximity of the promoter to the enhancer. Enhancers also function with transcription factors in order to initiate expression, much like silencers can with repressors.

In Prokaryotes and Eukaryotes

Prokaryotes

There are several differences in the regulation of metabolic control in eukaryotes and in prokaryotes. Prokaryotes vary the numbers of specific enzymes made in their cells in order to regulate gene expression, which is slow metabolic control, and also regulate enzymatic pathways through mechanisms such as feedback inhibition and allosteric regulation, which is rapid metabolic control.

The genes of prokaryotes are grouped together based on similar functions into units called operons which consist of a promoter and an operator. The operator is the binding site for the repressor and thus has a function equivalent to the silencer region in Eukaryotic DNA. When a repressor protein is bound to the operator, RNA polymerase cannot bind to the promoter to initiate the transcription of the operon.

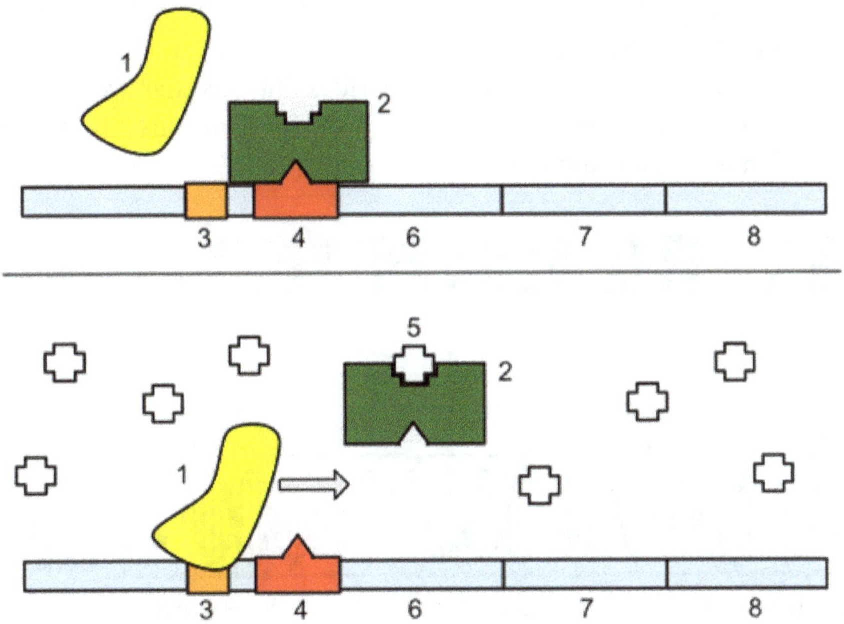

Fig: 1: RNA Polymerase, 2: Repressor (LacI), 3: Promoter, 4: Operator, 5: Lactose, 6: lacZ, 7: lacY, 8: lacA. Top: *Lac* Operon is Initially Repressed because Lactose is not Present to Inhibit the Repressor. Bottom: Repressor LacI is Inhibited because it Binds to Lactose and Transcription of the *Lac* Operon is Initiated for the Breakdown of Lactose

Repression of the Lac Operon

The *lac* operon in the prokaryote E. coli consists of genes that produce enzymes to break down lactose. Its operon is an example of a prokaryotic silencer. The three functional genes in this operon are lacZ, lacY, and lacA. The repressor gene, lacI, will produce the repressor protein LacI which is under allosteric regulation. These genes are activated by the presence of lactose in the cell which acts as an effector molecule that binds to LacI. When the repressor is bound to lactose, it will not bind to the operator, which allows RNA polymerase to bind to the promoter to initiate transcription of the operon. When the repressor's allosteric site is not bound to lactose, its active site will bind to the operator to prevent RNA polymerase from transcribing the genes of the *lac* operon.

Eukaryotes

Eukaryotes have a much larger genome and thus have different methods of gene regulation than in prokaryotes. All cells in a eukaryotic organism have the same DNA but are specified through differential gene expression, a phenomenon known as genetic totipotency. However, in order for a cell to express the genes for proper functioning, the genes must be closely regulated to express the correct properties. Genes in eukaryotes are controlled on the transcriptional, post-transcriptional, translational, and post-translational levels. On the transcriptional level, gene expression is regulated by altering transcription rates. Genes that encode proteins include exons which will encode the polypeptides, introns that are removed from mRNA before the translation of proteins, a transcriptional start site in which RNA polymerase binds, and a promoter.

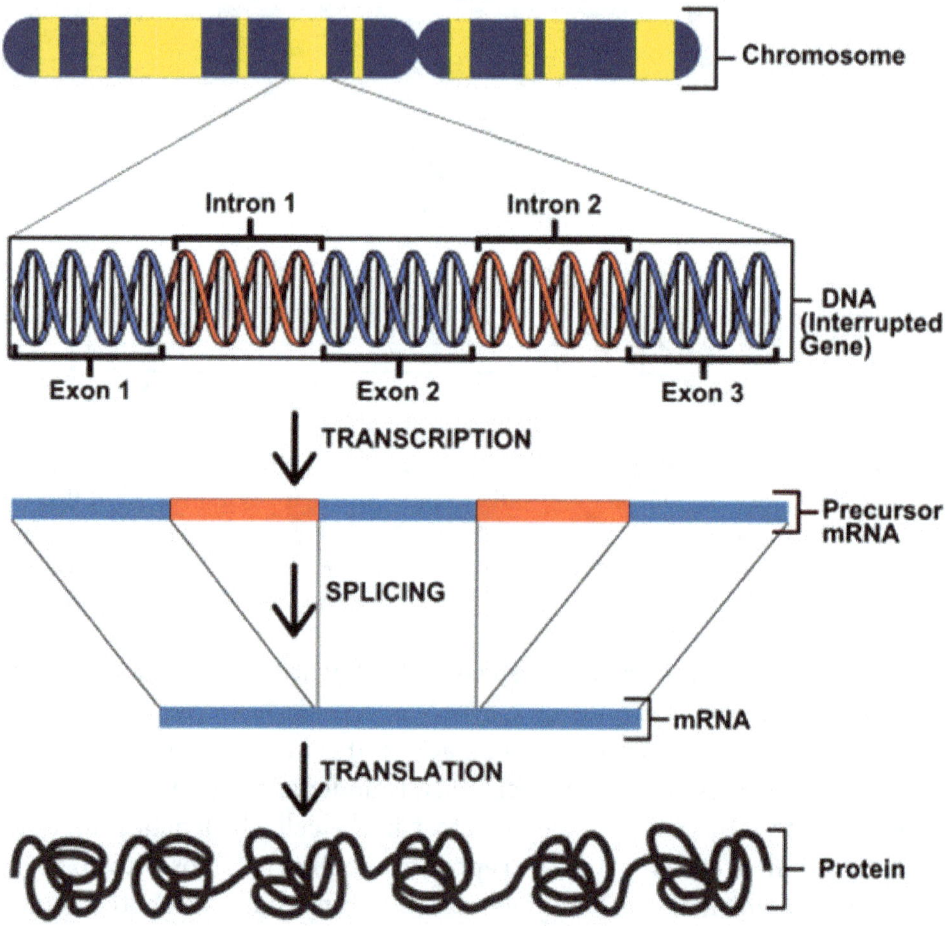

Fig: DNA is Transcribed into mRNA, Introns are Spliced during Post-Transcriptional Regulation, and the Remaining Exons Comprise the mRNA

Repression of the TATA Box

Eukaryotic genes contain an upstream promoter and a core promoter also referred to as a basal promoter. A common basal promoter is the TATAAAAAA sequence known as the TATA box.

The TATA box is a complex with several different proteins including transcription factor II D (TFIID) which includes the TATA-binding protein (TBP) that binds to the

TATA box along with 13 other proteins that bind to TBP. The TATA box binding proteins also include the transcription factor II B (TFIIB) which binds to both DNA and RNA polymerases.

Silencers in eukaryotes control gene expression on a transcriptional level in which the mRNA is not transcribed. These DNA sequences may act as either silencers or enhancers based on the transcription factor that binds to the sequence and binding of this sequence will prevent promoters such as the TATA box from binding to RNA polymerase. A repressor protein may have regions that bind to the DNA sequence as well as regions that bind to the transcription factors assembled at the promoter of the gene which would create a chromosome looping mechanism. Looping brings silencers in close proximity to the promoters to ensure that groups of proteins needed for optimal gene expression will work together.

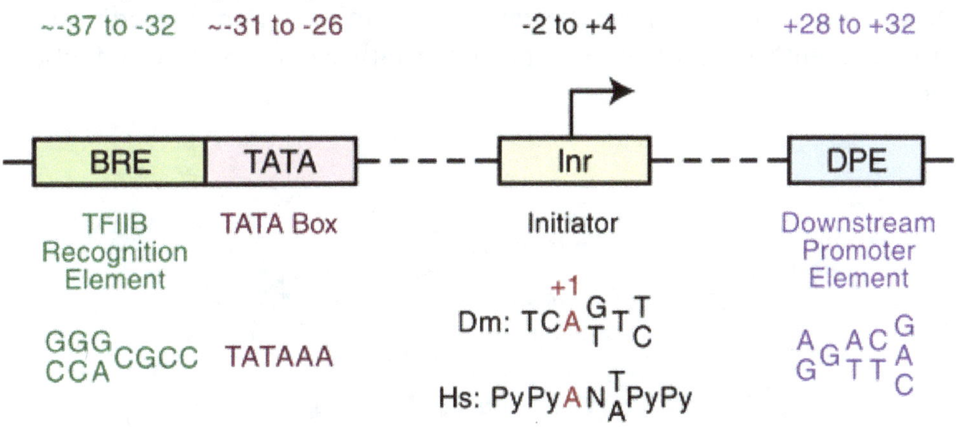

Fig: TATA Box, a Common Basal Promoter in Eukaryotes. The TATA Box is Grouped with the TFIIB and the Transcription Initiator Site and the Downstream Promoter Element are Located Several Base Pairs Away

Mutated Silencers, Hereditary Diseases, and their Effects

Genetic mutations occur when nucleotide sequences in an organism are altered. These mutations lead to not only observable phenotypic influences in an individual, but also alterations that are undetectable phenotypically. The sources for these mutations can be errors during replication, spontaneous mutations, and chemical and physical mutagens (UV and ionizing radiation, heat).

Silencers, being encoded in the genome, are susceptible to such alterations which, in many cases, can lead to severe phenotypical and functional abnormalities. In general terms, mutations in silencer elements or regions could lead to either the inhibition of the silencer's action or to the persisting repression of a necessary gene. This can then lead to the expression or suppression of an undesired phenotype which may affect the normal functionality of certain systems in the organism. Among the many silencer elements and proteins, REST/NSRF is an important silencer factor that has a variety of impacts, not only in neural aspects of development. In fact, in many cases, REST/NSRF acts in conjunction with RE-1/NRSE to repress and influence non-neuronal cells. Its effects range from frogs (*Xenopus laevis*) to humans, with innumerous effects in phenotype and also in development. In *Xenopus laevis*, REST/NRSF malfunction or damage has been associated to abnormal ectodermal patterning during development and significant consequences in neural tube, cranial ganglia, and eye development. In humans, a deficiency in the REST/NSRF silencer element has been correlated to Huntington's disease due to the decrease in the transcription of BDNF.

Furthermore, ongoing studies indicate that NRSE is involved in the regulation of the ANP gene, which when over expressed, can lead to ventricular hypertrophy. Mutations in the Polycomb-group (PcG) complexes also presented significant modifications in physiological systems of organisms. Hence, modification in silencer elements and sequences can result in either devastating or unnoticeable changes.

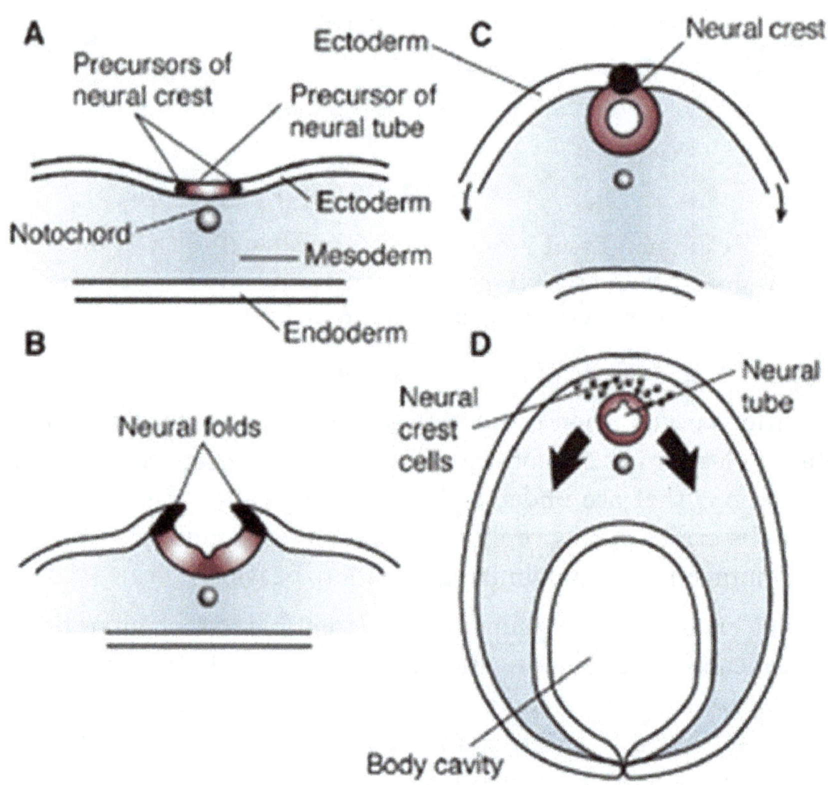

Fig: Proper neural folding.

Specialized cells called the notochord (A) induces ectoderm above it to become the primitive nervous system. (B) Neural tube forms (C) Gives rise to the brain and spinal cord. (D) Neural crest cells will migrate to different regions throughout the embryo to initiate development of glia, pigments, and other neural structures. Abnormal ectoderm patterning will cause abnormal and no neural folding.

REST/NRSF in Xenopus Laevis

The effects and influences of RE1/NRSE and REST/NRSF are significant in non-neuronal cells that require the repression or silencing of neuronal genes. These silencer elements also regulate the expression of genes that do not induce neuron-specific proteins and studies have shown the extensive impact these factors have in cellular processes. In Xenopus laevis, RE1/NRSE and REST/NRSF dysfunction or mutation demonstrated significant impact on neural tube, cranial ganglia, and eye development. All of these alterations can be traced to an improper patterning of the ectoderm during Xenopus development. Thus, a mutation or alteration in either the silencing region RE1/NRSE or silencer REST/NRSF factor can disrupt the proper differentiation and specification of the neuroepithelial domain and also hinder the formation of skin or ectoderm. The lack of these factors result in a decreased production of bone morphogenetic protein (BMP), which translates into a deficient development of the neural crest. Hence, the effects of NRSE and NRSF are of fundamental importance for neurogenesis of the developing embryo, and also in the early stages of ectodermal patterning. Ultimately, inadequate functioning of these factors can result in aberrant neural tube, cranial ganglia, and eye development in *Xenopus*.

REST/NSRF and Huntington's Disease

Huntington's disease (HD) is an inherited neurodegenerative disorder, with symptoms emerging during an individual's mid-adulthood. The most noticeable symptoms of this progressive disease are cognitive and motor impairments, as well as behavioral alterations. These impairments can develop into dementia, chorea, and eventually death. At the molecular level, HD results from a mutation in the huntingtin protein (Htt). More specifically, there is an abnormal repetition of a CAG sequence towards the 5'-end of the gene, which then leads to the development of a toxic polyglutamine (polyQ) stretch in the protein. The mutated Htt protein affects an individual's proper neural functions by inhibiting the action of REST/NRSF.

REST/NRSF is an important silencer element that binds to regulatory regions to control the expression of certain proteins involved in neural functions. The mechanistic actions of huntingtin are still not fully understood, but a correlation between Htt and REST/NRSF exists in HD development. By attaching to the REST/

NRSF, the mutated huntingtin protein inhibits the action of the silencer element, and retains it in the cytosol. Thus, REST/NRSF cannot enter the nucleus and bind to the 21 base-pair RE-1/NRSE regulatory element. An adequate repression of specific target genes are of fundamental importance, as many are involved in the proper development of neuronal receptors, neurotransmitters, synaptic vesicle proteins, and channel proteins.

A deficiency in the proper development of these proteins can cause the neural dysfunctions seen in Huntington's disease. In addition to the lack of repression due to the inactive REST/NRSF, mutated huntingtin protein can also decrease the transcription of the brain-derived neurotropic factor (BDNF) gene. BDNF influences the survival and development of neurons in the central nervous system as well as the peripheral nervous system. This abnormal repression occurs when the RE1/NRSE region within the BDNF promoter region is activated by the binding of REST/NRSF, which leads to the lack of transcription of the BDNF gene. Hence, the anomalous repression of the BDNF protein suggests a significant impact in Huntington's disease.

Current Research on REST/NRSF and Ventricular Hypertrophy in Mammals

REST/NRSF in conjunction with RE1/NRSE also acts outside the nervous system as regulators and repressors. Current research has linked RE1/NRSE activity with the regulation of the expression of the atrial natriuretic peptide (*ANP*) gene. An NRSE regulatory region is present in the 3' untranslated region of the *ANP* gene and acts as a mediator for its appropriate expression. The protein encoded by the *ANP* gene is important during embryonic development for the maturation and development of cardiac myocytes.

However, during early childhood and throughout adulthood, ANP expression is suppressed or kept to a minimum in the ventricle. Thus, an abnormal induction of the *ANP* gene can lead to ventricular hypertrophy and severe cardiac consequences. In order to maintain the repression of the gene, NRSF (neuron-restrictive silencer factor) or REST binds to the NRSE region in the 3'untranslated region of the *ANP* gene. Furthermore, the NRSF-NRSE complex recruits a transcriptional corepressor known as mSin3. This leads to the activity of histone deacetylase in the region and the repression of the gene.

Therefore, studies have revealed the correlation between REST/NRSF and RE1/NRSE in regulating the *ANP* gene expression in ventricular myocytes. A mutation in either the NRSF or NRSE can lead to an undesirable development of ventricular myocytes, due to lack of repression, which can then cause ventricular hypertrophy.

Left ventricular hypertrophy, for example, increases an individual's chance of sudden death due to a ventricular arrhythmia resulting from the increased ventricular mass. In addition to the influence on the *ANP* gene, the NRSE sequence regulates other cardiac embryonic genes, such as brain natriuretic peptide BNP, skeletal α-actin, and Na, K - ATPase α3 subunit.

Hence, the regulatory activity of both NRSE and NRSF in mammals prevents not only neural dysfunctions but also physiological and phenotypical abnormalities in other non-neuronal regions of the body.

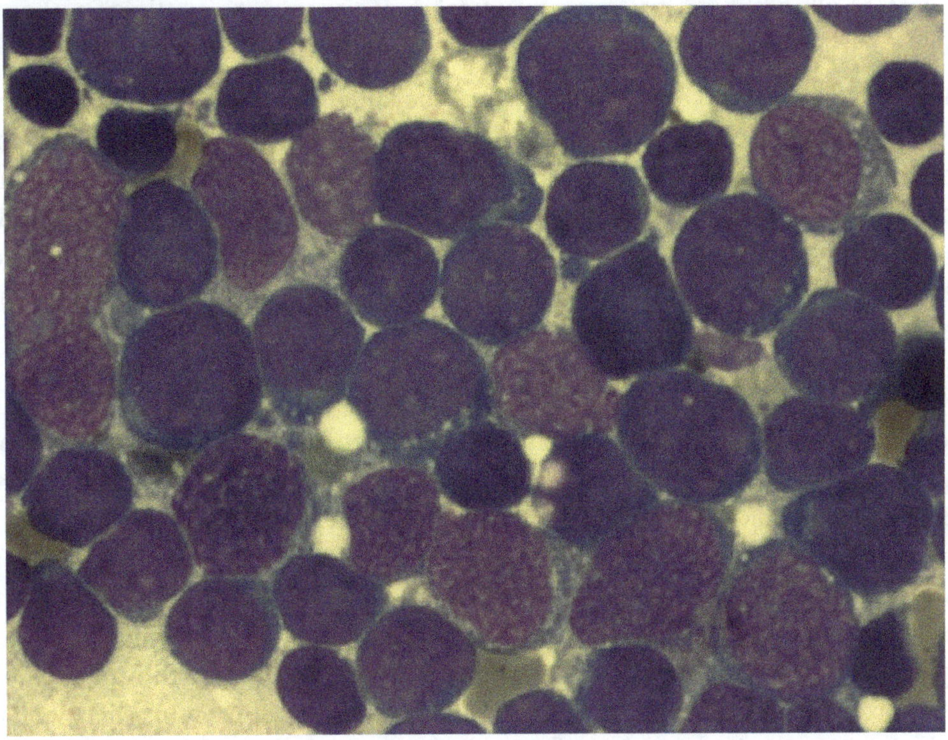

Fig: Bone Marrow of a Patient with Acute Lymphoblastic Leukemia

Mutations in Polycomb-Group Response Elements (PREs)

The Polycomb-group (PcG) regulatory complexes are known for their influence in the epigenetic regulation of stem cells, especially in hematopoietic stem cells.

The Polycomb Repressive Complex 1 (PRC 1) is directly involved in the process of hematopoiesis, and functions together with, for example, the PcG gene "Bmi1". Studies in mice indicate that organisms with mutated "Bmi1" demonstrate deficient mitochondrial functioning, and also hindered the ability of hematopoietic cells to self-renew.

Likewise, mutations in PRC2 genes were related to hematological conditions such as acute lymphoblastic leukemia (ALL), which is a form of leukemia. Hence, Polycomb-group genes and proteins are involved in the proper maintenance of hematopoiesis in the body.

5.5 TRANSCRIPTION FACTORS: PROTEINS THAT CONTROL GENE EXPRESSION

Transcription factors are proteins that bind to specific DNA sequences, typically within promoters, enhancers, or silencers, and regulate the transcription of eukaryotic genes. They act as molecular switches, turning genes on or off in response to various signals. Transcription factors often contain specific domains for DNA binding and activation or repression of transcription, allowing them to interact with the transcription machinery and modulate gene expression directly.

Transcription factors are essential for coordinated gene expression during development, as they activate or repress genes in a precise spatial and temporal pattern. They are also involved in responding to extracellular signals, such as hormones, which trigger transcription factors to bind to target genes and alter their expression. By integrating various signals, transcription factors allow cells to adapt their behavior to the organism's needs, playing a central role in processes like growth, differentiation, and immune responses.

5.6 RNA SPLICING AND ALTERNATIVE SPLICING

In eukaryotes, the initial RNA transcript produced during transcription, known as pre-mRNA, undergoes a process called RNA splicing before it becomes mature messenger RNA (mRNA). During splicing, introns are removed, and exons are joined together to form a continuous coding sequence. This process is catalyzed by a large complex known as the spliceosome, which recognizes specific sequences at the exon-intron boundaries. Alternative splicing, a variation of RNA splicing, allows a single gene to produce multiple mRNA isoforms by selectively including or excluding certain exons. This process greatly increases the diversity of proteins that a cell can produce from a limited number of genes. Alternative splicing plays a crucial role in tissue specificity and cellular differentiation, as different cell types may express different spliced variants of a gene, resulting in distinct protein functions.

5.7 POLYADENYLATION AND THE POLY-A TAIL

Another important post-transcriptional modification in eukaryotic gene expression is polyadenylation, the addition of a poly-A tail to the 3' end of the pre-mRNA. This poly-A tail consists of a long sequence of adenine nucleotides, which protects the mRNA from degradation, assists in its export from the nucleus, and facilitates translation initiation in the cytoplasm. The polyadenylation signal sequence, found near the end of the gene, directs the addition of the poly-A tail. The poly-A tail also contributes to the regulation of mRNA stability. mRNAs with longer poly-A tails tend to be more stable and have a longer half-life, which allows for prolonged protein synthesis. Polyadenylation is therefore essential for controlling gene expression levels, as it influences the availability of mRNA for translation.

5.8 GENE ORGANIZATION: SINGLE COPY GENES, GENE FAMILIES, AND REPETITIVE DNA

Eukaryotic genomes contain various types of gene organization. Single-copy genes are unique sequences that exist in only one location within the genome and typically code for essential proteins. In contrast, gene families are groups of related genes with similar sequences and functions, arising from duplication events. Members of a gene family may perform similar roles or have specialized functions in different tissues or stages of development. Repetitive DNA is another characteristic of eukaryotic genomes, consisting of sequences that are repeated multiple times. These sequences include satellite DNA, mini-satellites, and micro-satellites, which are found primarily in non-coding regions and may play structural roles, such as in centromeres and telomeres. Repetitive DNA may also have a regulatory role in gene expression, though its functions are still being explored.

5.9 EPIGENETIC REGULATION OF GENE EXPRESSION

Epigenetic modifications refer to chemical changes to the DNA or associated histone proteins that affect gene expression without altering the DNA sequence itself. Common epigenetic modifications include DNA methylation and histone modification. DNA methylation, the addition of a methyl group to cytosine bases, typically silences gene expression, while histone modifications, such as acetylation, can either activate or repress transcription depending on the type and location of the modification.

Epigenetic modifications are essential for regulating gene expression in response to environmental cues and during development. They allow cells to "remember" which genes should be active or inactive in specific cell types, contributing to cellular memory and stable gene expression patterns. Epigenetics also plays a role in phenomena such as X-chromosome inactivation and genomic imprinting.

5.10 CONCLUSION

The fundamental features of eukaryotic genes—such as the presence of introns and exons, complex regulatory elements, transcription factors, and post-transcriptional modifications—reflect the sophisticated mechanisms needed for gene expression in multicellular organisms. These features allow for a high degree of flexibility and specificity in gene regulation, enabling cells to respond to various developmental and environmental signals. Additionally, the presence of gene families, repetitive DNA, and epigenetic modifications add layers of complexity to eukaryotic gene function, allowing organisms to control gene expression in nuanced and adaptable ways. Together, these features highlight the complexity and diversity of eukaryotic genomes, underscoring the remarkable adaptability and specialization seen in eukaryotic life.

CHAPTER 6

A NEW TOOLBOX FOR RECOMBINANT DNA

6.1 INTRODUCTION TO RECOMBINANT DNA TECHNOLOGY

Recombinant DNA (rDNA) technology revolutionized genetics by enabling scientists to manipulate DNA sequences in ways previously unimaginable. By combining DNA from different organisms, rDNA allows for the synthesis of new genetic constructs that can be used for gene studies, protein production, genetic therapies, and the improvement of crops and livestock. In recent years, the toolbox for creating recombinant DNA has expanded significantly, with the development of innovative molecular tools and techniques. These advancements have made it possible to manipulate DNA with higher precision, efficiency, and versatility, bringing forth new possibilities in genetic engineering, medicine, and biotechnology.

6.2 TRADITIONAL TOOLS OF RECOMBINANT DNA TECHNOLOGY

Before the advent of the latest tools, the creation of recombinant DNA relied on techniques such as restriction enzymes, ligases, and plasmids. Restriction enzymes, often called molecular scissors, cut DNA at specific sequences, enabling the precise cleavage needed for inserting a gene of interest into a plasmid. DNA ligase then joins the cut DNA fragments, securing the foreign DNA into the plasmid. Plasmids, small circular DNA molecules, serve as vectors to transfer the inserted gene into host cells. These tools laid the foundation of recombinant DNA technology and allowed scientists to clone, express, and study genes.

However, traditional tools had limitations. Restriction enzymes, for instance, recognize only specific DNA sequences, which limited the flexibility of genetic manipulation. Moreover, early techniques often lacked precision in controlling the insertion of genes within the host genome, leading to inconsistent results. Recognizing these challenges, researchers sought more sophisticated methods to enhance efficiency and expand the scope of recombinant DNA applications.

6.3 THE RISE OF CRISPR-CAS9: A REVOLUTIONARY GENE EDITING TOOL

CRISPR is a family of DNA sequences found in the genomes of prokaryotic organisms such as bacteria and archaea. Each sequence within an individual prokaryotic cell is derived from a DNA fragment of a bacteriophage that had previously infected the prokaryote or one of its ancestors. These sequences are used to detect and destroy DNA from similar bacteriophages during subsequent infections. Hence these sequences play a key role in the antiviral (i.e., anti-phage) defense system of prokaryotes and provide a form of heritable, acquired immunity. CRISPR is found in approximately 50% of sequenced bacterial genomes and nearly 90% of sequenced archaea.

Cas9 (or "CRISPR-associated protein 9") is an enzyme that uses CRISPR sequences as a guide to recognize and open up specific strands of DNA that are complementary to the CRISPR sequence. Cas9 enzymes together with CRISPR sequences form the basis of a technology known as CRISPR-Cas9 that can be used to edit genes within living organisms. This editing process has a wide variety of applications including basic biological research, development of biotechnological products, and treatment of diseases. The development of the CRISPR-Cas9 genome editing technique was recognized by the Nobel Prize in Chemistry in 2020 awarded to Emmanuelle Charpentier and Jennifer Doudna.

6.3.1 History

Repeated Sequences

The discovery of clustered DNA repeats took place independently in three parts of the world. The first description of what would later be called CRISPR is from Osaka University researcher Yoshizumi Ishino and his colleagues in 1987. They accidentally cloned part of a CRISPR sequence together with the "*iap*" *gene (isozyme conversion of alkaline phosphatase)* from their target genome, that of *Escherichia coli*. The organization of the repeats was unusual. Repeated sequences are typically arranged consecutively, without interspersing different sequences. They did not know the function of the interrupted clustered repeats.

In 1993, researchers of *Mycobacterium tuberculosis* in the Netherlands published two articles about a cluster of interrupted direct repeats (DR) in that bacterium. They recognized the diversity of the sequences that intervened in the direct repeats among different strains of *M. tuberculosis* and used this property to design a typing method called *spoligotyping*, still in use today.

Francisco Mojica at the University of Alicante in Spain studied the function of repeats in the archaeal species *Haloferax* and *Haloarcula*. Mojica's supervisor surmised that the clustered repeats had a role in correctly segregating replicated DNA into daughter cells during cell division, because plasmids and chromosomes with identical repeat arrays could not coexist in *Haloferax volcanii*. Transcription of the interrupted repeats was also noted for the first time; this was the first full characterization of CRISPR. By 2000, Mojica and his students, after an automated search of published genomes, identified interrupted repeats in 20 species of microbes as belonging to the same family. Because those sequences were interspaced, Mojica initially called these sequences "short regularly spaced repeats" (SRSR). In 2001, Mojica and Ruud Jansen, who were searching for an additional interrupted repeats, proposed the acronym CRISPR (Clustered Regularly Interspaced Short Palindromic Repeats) to unify the numerous acronyms used to describe these sequences.

In 2002, Tang, et al. showed evidence that CRISPR repeat regions from the genome of *Archaeoglobus fulgidus* were transcribed into long RNA molecules subsequently processed into unit-length small RNAs, plus some longer forms of 2, 3, or more spacer-repeat units.

In 2005, yogurt researcher Rodolphe Barrangou discovered that *Streptococcus thermophilus*, after iterative phage infection challenges, develops increased phage resistance due to the incorporation of additional CRISPR spacer sequences. Barrangou's employer, the Danish food company Danisco, then developed phage-resistant *S. thermophilus* strains for yogurt production. Danisco was later bought by DuPont, which owns about 50 percent of the global dairy culture market, and the technology spread widely.

CRISPR-Associated Systems

A major advance in understanding CRISPR came with Jansen's observation that the prokaryote repeat cluster was accompanied by four homologous genes that make up CRISPR-associated systems, *cas* 1–4. The Cas proteins showed helicase and nuclease motifs, suggesting a role in the dynamic structure of the CRISPR loci. In this publication, the acronym CRISPR was used as the universal name of this pattern, but its function remained enigmatic.

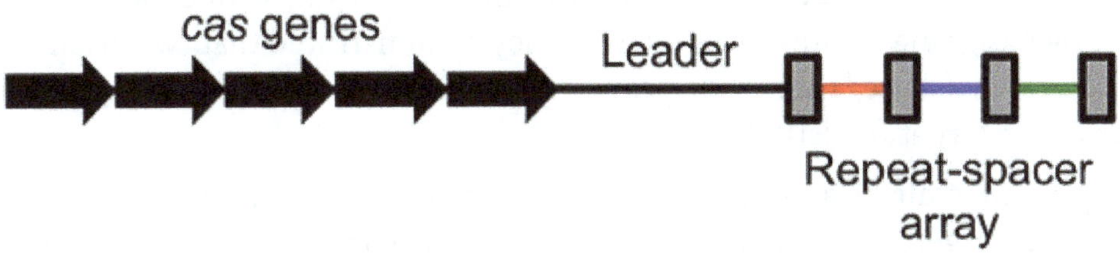

Fig: Simplified Diagram of a CRISPR Locus

The three major components of a CRISPR locus are shown: *cas* genes, a leader sequence, and a repeat-spacer array. Repeats are shown as gray boxes and spacers are colored bars. The arrangement of the three components is not always as shown. In addition, several CRISPRs with similar sequences can be present in a single genome, only one of which is associated with *cas* genes.

In 2005, three independent research groups showed that some CRISPR spacers are derived from phage DNA and extrachromosomal DNA such as plasmids. In effect, the spacers are fragments of DNA gathered from viruses that previously attacked the cell. The source of the spacers was a sign that the CRISPR-*cas* system could have a role in adaptive immunity in bacteria. All three studies proposing this idea were initially rejected by high-profile journals, but eventually appeared in other journals.

The first publication proposing a role of CRISPR-Cas in microbial immunity, by Mojica and collaborators at the University of Alicante, predicted a role for the RNA transcript of spacers on target recognition in a mechanism that could be analogous to the RNA interference system used by eukaryotic cells. Koonin and colleagues extended this RNA interference hypothesis by proposing mechanisms of action for the different CRISPR-Cas subtypes according to the predicted function of their proteins.

Experimental work by several groups revealed the basic mechanisms of CRISPR-Cas immunity. In 2007, the first experimental evidence that CRISPR was an adaptive immune system was published.

A CRISPR region in *Streptococcus thermophilus* acquired spacers from the DNA of an infecting bacteriophage. The researchers manipulated the resistance of *S. thermophilus* to different types of phages by adding and deleting spacers whose sequence matched those found in the tested phages. In 2008, Brouns and Van der Oost identified a complex of Cas proteins called Cascade, that in *E. coli* cut the CRISPR RNA precursor within the repeats into mature spacer-containing RNA molecules called CRISPR RNA (crRNA), which remained bound to the protein complex.

Moreover, it was found that Cascade, crRNA and a helicase/nuclease (Cas3) were required to provide a bacterial host with immunity against infection by a DNA virus. By designing an anti-virus CRISPR, they demonstrated that two orientations of the crRNA (sense/antisense) provided immunity, indicating that the crRNA guides were targeting dsDNA.

That year Marraffini and Sontheimer confirmed that a CRISPR sequence of *S. epidermidis* targeted DNA and not RNA to prevent conjugation. This finding was at odds with the proposed RNA-interference-like mechanism of CRISPR-Cas immunity, although a CRISPR-Cas system that targets foreign RNA was later found in *Pyrococcus furiosus*. A 2010 study showed that CRISPR-Cas cuts strands of both phage and plasmid DNA in *S. thermophilus*.

Cas9

A simpler CRISPR system from *Streptococcus pyogenes* relies on the protein Cas9. The Cas9 endonuclease is a four-component system that includes two small molecules: crRNA and trans-activating CRISPR RNA (tracrRNA). In 2012, Jennifer Doudna and Emmanuelle Charpentier re-engineered the Cas9 endonuclease into a more manageable two-component system by fusing the two RNA molecules into a "single-guide RNA" that, when combined with Cas9, could find and cut the DNA target specified by the guide RNA.

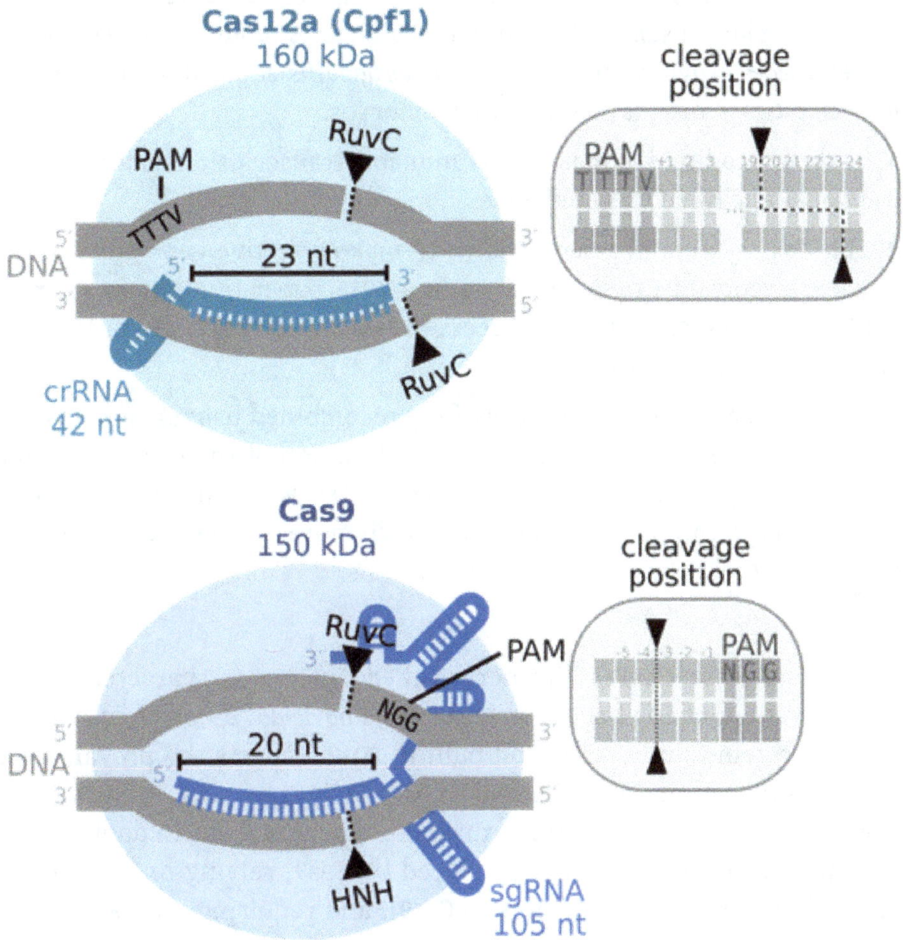

Fig: A Diagram of the CRISPR Nucleases Cas12a and Cas9 with the Position of DNA Cleavage shown Relative to their PAM Sequences in a Zoom-In

This contribution was so significant that it was recognized by the Nobel Prize in Chemistry in 2020. By manipulating the nucleotide sequence of the guide RNA, the artificial Cas9 system could be programmed to target any DNA sequence for separation.

Another collaboration comprising Virginijus Šikšnys, Gasiūnas, Barrangou, and Horvath showed that Cas9 from the *S. thermophilus* CRISPR system can also be reprogrammed to target a site of their choosing by changing the sequence of its crRNA. These advances fueled efforts to edit genomes with the modified CRISPR-Cas9 system.

Groups led by Feng Zhang and George Church simultaneously published descriptions of genome editing in human cell cultures using CRISPR-Cas9 for the first time. It has since been used in a wide range of organisms, including baker's yeast (*Saccharomyces cerevisiae*), the opportunistic pathogen *Candida albicans*, zebrafish (*Danio rerio*), fruit flies (*Drosophila melanogaster*), ants (*Harpegnathos saltator* and *Ooceraea biroi*), mosquitoes (*Aedes aegypti*), nematodes (*Caenorhabditis elegans*), plants, mice (*Mus musculus domesticus)*, monkeys and human embryos.

CRISPR has been modified to make programmable transcription factors that allows activation or silencing of targeted genes.

The CRISPR-Cas9 system has been shown to make effective gene edits in Human tripronuclear zygotes, as first described in a 2015 paper by Chinese scientists P. Liang and Y. Xu. The system made a successful cleavage of mutant Beta-Hemoglobin (HBB) in 28 out of 54 embryos.

Four out of the 28 embryos were successfully recombined using a donor template. The scientists showed that during DNA recombination of the cleaved strand, the homologous endogenous sequence HBD competes with the exogenous donor template. DNA repair in human embryos is much more complicated and particular than in derived stem cells.

Cas12a

In 2015, the nuclease Cas12a (formerly called Cpf1) was characterized in the CRISPR-Cpf1 system of the bacterium *Francisella novicida*. Its original name, from a TIGRFAMs protein family definition built in 2012, reflects the prevalence of its CRISPR-Cas subtype in the *Prevotella* and *Francisella* lineages. Cas12a showed several key differences from Cas9 including: causing a 'staggered' cut in double stranded DNA as opposed to the 'blunt' cut produced by Cas9, relying on a 'T rich' PAM (providing alternative targeting sites to Cas9), and requiring only a CRISPR RNA (crRNA) for successful targeting. By contrast, Cas9 requires both crRNA and a trans-activating crRNA (tracrRNA).

These differences may give Cas12a some advantages over Cas9. For example, Cas12a's small crRNAs are ideal for multiplexed genome editing, as more of them can be packaged in one vector than can Cas9's sgRNAs.

The sticky 5' overhangs left by Cas12a can also be used for DNA assembly that is much more target-specific than traditional restriction enzyme cloning. Finally, Cas12a cleaves DNA 18–23 base pairs downstream from the PAM site. This means there is no disruption to the recognition sequence after repair, and so Cas12a enables multiple rounds of DNA cleavage.

By contrast, since Cas9 cuts only 3 base pairs upstream of the PAM site, the NHEJ pathway results in indel mutations that destroy the recognition sequence, thereby preventing further rounds of cutting. In theory, repeated rounds of DNA cleavage should cause an increased opportunity for the desired genomic editing to occur.

A distinctive feature of Cas12a, as compared to Cas9, is that after cutting its target, Cas12a remains bound to the target and then cleaves other ssDNA molecules non-discriminately. This property is called "collateral cleavage" or "trans-cleavage" activity and has been exploited for the development of various diagnostic technologies.

Cas13

In 2016, the nuclease Cas13a (formerly known as C2c2) from the bacterium *Leptotrichia shahii* was characterized. Cas13 is an RNA-guided RNA endonuclease, which means that it does not cleave DNA, but only single-stranded RNA. Cas13 is guided by its crRNA to a ssRNA target and binds and cleaves the target. Similar to Cas12a, the Cas13 remains bound to the target and then cleaves other ssRNA molecules non-discriminately.

This collateral cleavage property has been exploited for the development of various diagnostic technologies.

In 2021, Dr. Hui Yang characterized novel miniature Cas13 protein (mCas13) variants, Cas13X and Cas13Y. Using a small portion of N gene sequence from SARS-CoV-2 as a target in characterization of mCas13, revealed the sensitivity and specificity of mCas13 coupled with RT-LAMP for detection of SARS-CoV-2 in both synthetic and clinical samples over other available standard tests like RT-qPCR (1 copy/μL).

6.3.2 Locus Structure

Repeats and Spacers

The CRISPR array is made up of an AT-rich leader sequence followed by short repeats that are separated by unique spacers. CRISPR repeats typically range in size from 28 to 37 base pairs (bps), though there can be as few as 23 bp and as many as 55 bp. Some show dyad symmetry, implying the formation of a secondary structure such as a stem-loop ('hairpin') in the RNA, while others are designed to be unstructured.

The size of spacers in different CRISPR arrays is typically 32 to 38 bp (range 21 to 72 bp). New spacers can appear rapidly as part of the immune response to phage infection. There are usually fewer than 50 units of the repeat-spacer sequence in a CRISPR array.

CRISPR RNA structures

Fig: CRISPR-DR2: Secondary Structure taken from the Rfam Database

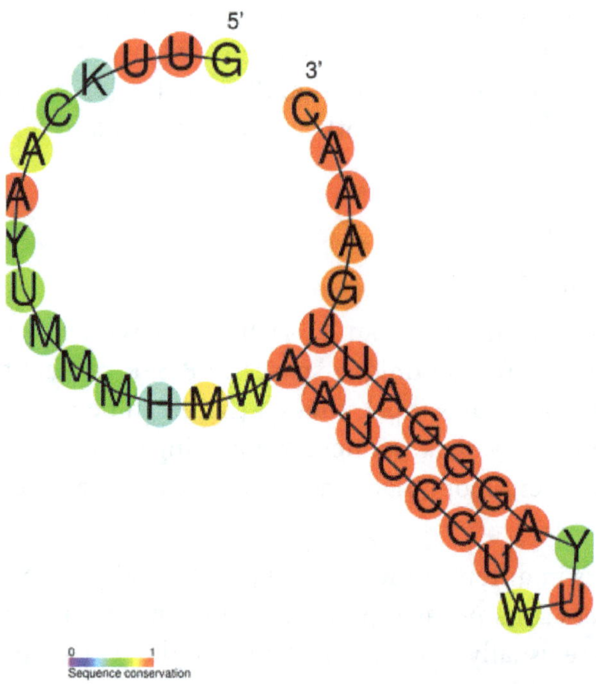

Fig: CRISPR-DR5: Secondary Structure taken from the Rfam Database

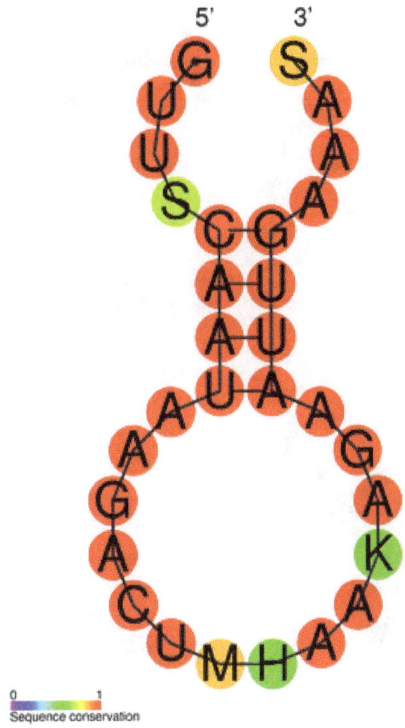

Fig: CRISPR-DR6: Secondary Structure taken from the Rfam Database

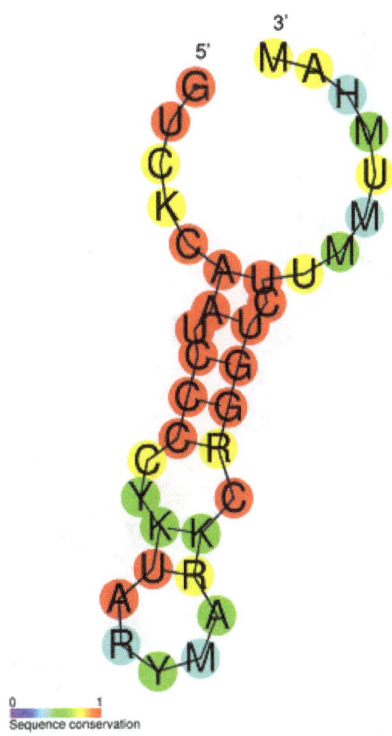

Fig: CRISPR-DR8: Secondary Structure Taken from the Rfam Database

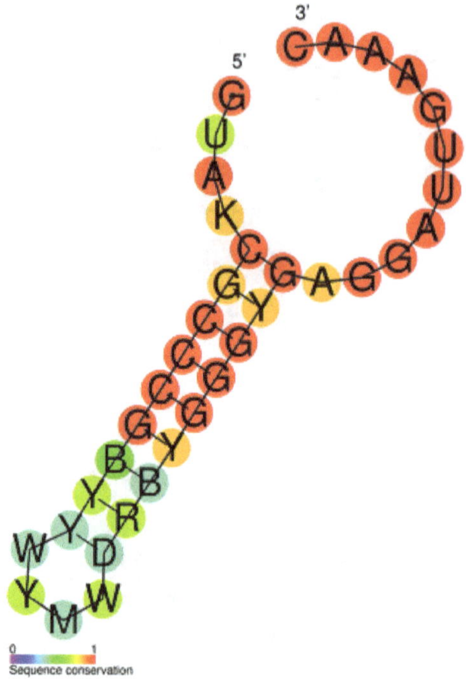

Fig: CRISPR-DR9: Secondary Structure taken from the Rfam Database

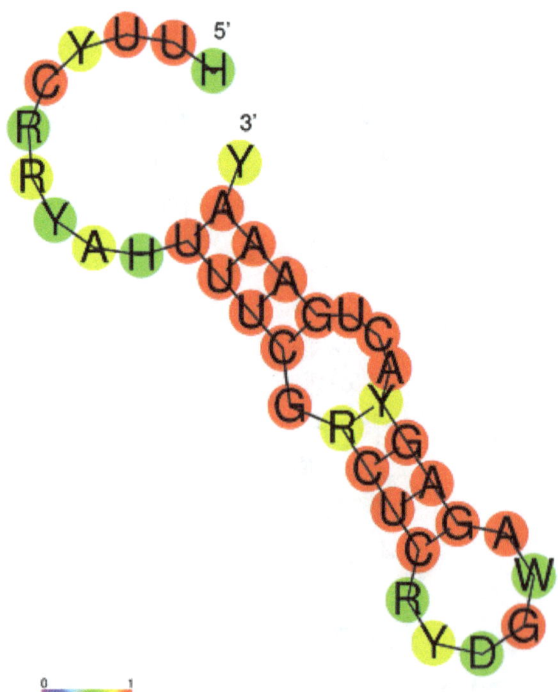

Fig: CRISPR-DR19: Secondary Structure taken from the Rfam Database

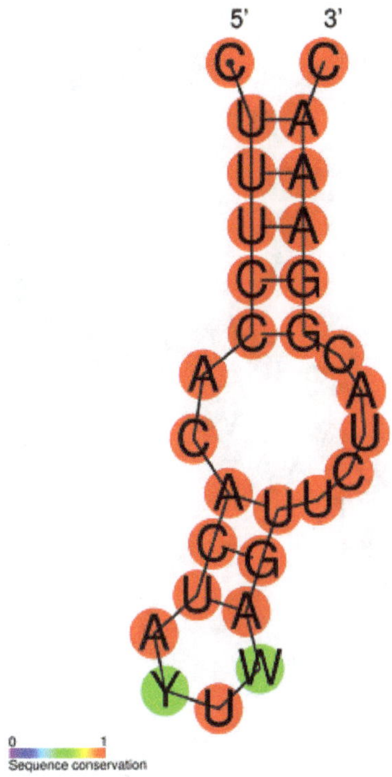

Fig: CRISPR-DR41: Secondary Structure taken from the Rfam Database

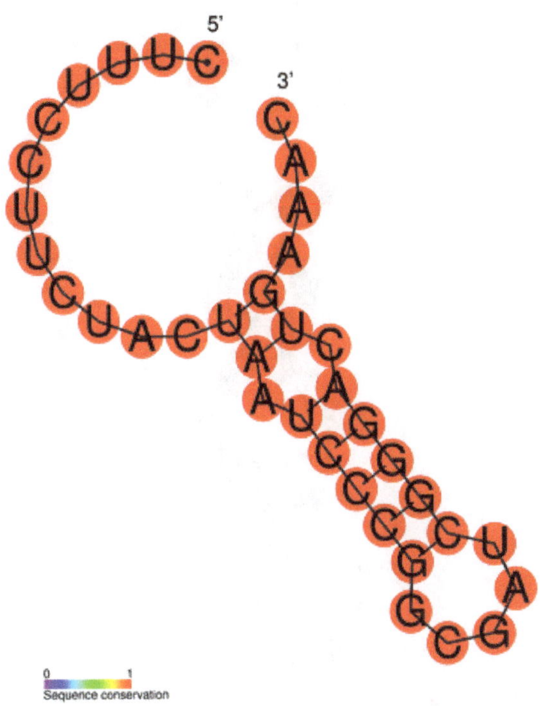

Fig: CRISPR-DR52: Secondary Structure taken from the Rfam Database

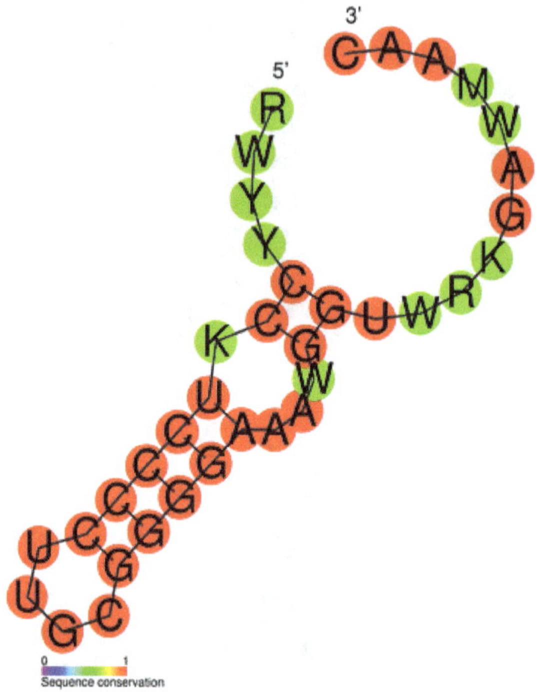

Fig: CRISPR-DR57: Secondary Structure taken from the Rfam Database

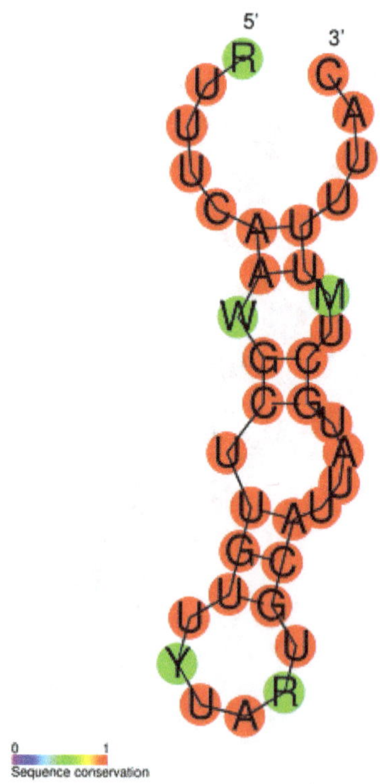

Fig: CRISPR-DR65: Secondary Structure taken from the Rfam Database

Cas Genes and CRISPR Subtypes

Small clusters of *cas* genes are often located next to CRISPR repeat-spacer arrays. Collectively the 93 *cas* genes are grouped into 35 families based on sequence similarity of the encoded proteins. 11 of the 35 families form the *cas* core, which includes the protein families Cas1 through Cas9. A complete CRISPR-Cas locus has at least one gene belonging to the *cas* core.

CRISPR-Cas systems fall into two classes. Class 1 systems use a complex of multiple Cas proteins to degrade foreign nucleic acids. Class 2 systems use a single large Cas protein for the same purpose. Class 1 is divided into types I, III, and IV; class 2 is divided into types II, V, and VI. The 6 system types are divided into 33 subtypes. Each type and most subtypes are characterized by a "signature gene" found almost exclusively in the category. Classification is also based on the complement of *cas* genes that are present. Most CRISPR-Cas systems have a Cas1 protein. The phylogeny of Cas1 proteins generally agrees with the classification system, but exceptions exist due to module shuffling. Many organisms contain multiple CRISPR-Cas systems suggesting that they are compatible and may share components. The sporadic distribution of the CRISPR-Cas subtypes suggests that the CRISPR-Cas system is subject to horizontal gene transfer during microbial evolution.

6.3.3 Mechanism

CRISPR-Cas immunity is a natural process of bacteria and archaea. CRISPR-Cas prevents bacteriophage infection, conjugation and natural transformation by degrading foreign nucleic acids that enter the cell.

Spacer Acquisition

When a microbe is invaded by a bacteriophage, the first stage of the immune response is to capture phage DNA and insert it into a CRISPR locus in the form of a spacer. Cas1 and Cas2 are found in both types of CRISPR-Cas immune systems, which indicates that they are involved in spacer acquisition. Mutation studies confirmed this hypothesis, showing that removal of Cas1 or Cas2 stopped spacer acquisition, without affecting CRISPR immune response.

Multiple Cas1 proteins have been characterised and their structures resolved. Cas1 proteins have diverse amino acid sequences. However, their crystal structures are similar and all purified Cas1 proteins are metal-dependent nucleases/integrases that bind to DNA in a sequence-independent manner. Representative Cas2 proteins have been characterised and possess either (single strand) ssRNA- or (double strand) dsDNA- specific endoribonuclease activity.

In the I-E system of *E. coli* Cas1 and Cas2 form a complex where a Cas2 dimer bridges two Cas1 dimers. In this complex Cas2 performs a non-enzymatic scaffolding role, binding double-stranded fragments of invading DNA, while Cas1 binds the single-stranded flanks of the DNA and catalyses their integration into CRISPR arrays. New spacers are usually added at the beginning of the CRISPR next to the leader sequence creating a chronological record of viral infections.

In *E. coli* a histone like protein called integration host factor (IHF), which binds to the leader sequence, is responsible for the accuracy of this integration. IHF also enhances integration efficiency in the type I-F system of *Pectobacterium atrosepticum*. but in other systems, different host factors may be required

Protospacer Adjacent Motifs (PAM)

Bioinformatic analysis of regions of phage genomes that were excised as spacers (termed protospacers) revealed that they were not randomly selected but instead were found adjacent to short (3–5 bp) DNA sequences termed protospacer adjacent motifs (PAM). Analysis of CRISPR-Cas systems showed PAMs to be important for type I and type II, but not type III systems during acquisition. In type I and type II systems, protospacers are excised at positions adjacent to a PAM sequence, with the other end of the spacer cut using a ruler mechanism, thus maintaining the regularity of the spacer size in the CRISPR array. The conservation of the PAM sequence differs between CRISPR-Cas systems and appears to be evolutionarily linked to Cas1 and the leader sequence.

New spacers are added to a CRISPR array in a directional manner, occurring preferentially, but not exclusively, adjacent to the leader sequence. Analysis of the type I-E system from *E. coli* demonstrated that the first direct repeat adjacent to the leader sequence is copied, with the newly acquired spacer inserted between the first and second direct repeats.

The PAM sequence appears to be important during spacer insertion in type I-E systems. That sequence contains a strongly conserved final nucleotide (nt) adjacent to the first nt of the protospacer.

This nt becomes the final base in the first direct repeat. This suggests that the spacer acquisition machinery generates single-stranded overhangs in the second-to-last position of the direct repeat and in the PAM during spacer insertion.

However, not all CRISPR-Cas systems appear to share this mechanism as PAMs in other organisms do not show the same level of conservation in the final position. It is likely that in those systems, a blunt end is generated at the very end of the direct repeat and the protospacer during acquisition.

Insertion Variants

Analysis of *Sulfolobus solfataricus* CRISPRs revealed further complexities to the canonical model of spacer insertion, as one of its six CRISPR loci inserted new spacers randomly throughout its CRISPR array, as opposed to inserting closest to the leader sequence.

Multiple CRISPRs contain many spacers to the same phage. The mechanism that causes this phenomenon was discovered in the type I-E system of *E. coli*. A significant enhancement in spacer acquisition was detected where spacers already target the phage, even mismatches to the protospacer. This 'priming' requires the Cas proteins involved in both acquisition and interference to interact with each other. Newly acquired spacers that result from the priming mechanism are always found on the same strand as the priming spacer. This observation led to the hypothesis that the acquisition machinery slides along the foreign DNA after priming to find a new protospacer.

Biogenesis

CRISPR-RNA (crRNA), which later guides the Cas nuclease to the target during the interference step, must be generated from the CRISPR sequence. The crRNA is initially transcribed as part of a single long transcript encompassing much of the CRISPR array.

This transcript is then cleaved by Cas proteins to form crRNAs. The mechanism to produce crRNAs differs among CRISPR-Cas systems. In type I-E and type I-F systems, the proteins Cas6e and Cas6f respectively, recognise stem-loops created by the pairing of identical repeats that flank the crRNA. These Cas proteins cleave the longer transcript at the edge of the paired region, leaving a single crRNA along with a small remnant of the paired repeat region.

Type III systems also use Cas6, however, their repeats do not produce stem-loops. Cleavage instead occurs by the longer transcript wrapping around the Cas6 to allow cleavage just upstream of the repeat sequence.

Type II systems lack the Cas6 gene and instead utilize RNaseIII for cleavage. Functional type II systems encode an extra small RNA that is complementary to the repeat sequence, known as a trans-activating crRNA (tracrRNA).

Transcription of the tracrRNA and the primary CRISPR transcript results in base pairing and the formation of dsRNA at the repeat sequence, which is subsequently targeted by RNaseIII to produce crRNAs. Unlike the other two systems, the crRNA does not contain the full spacer, which is instead truncated at one end.

CrRNAs associate with Cas proteins to form ribonucleotide complexes that recognize foreign nucleic acids. CrRNAs show no preference between the coding and non-coding strands, which is indicative of an RNA-guided DNA-targeting system. The type I-E complex (commonly referred to as Cascade) requires five Cas proteins bound to a single crRNA.

Interference

During the interference stage in type I systems, the PAM sequence is recognized on the crRNA-complementary strand and is required along with crRNA annealing. In type I systems correct base pairing between the crRNA and the protospacer signals a conformational change in Cascade that recruits Cas3 for DNA degradation.

Type II systems rely on a single multifunctional protein, Cas9, for the interference step. Cas9 requires both the crRNA and the tracrRNA to function and cleave DNA using its dual HNH and RuvC/RNaseH-like endonuclease domains. Basepairing between the PAM and the phage genome is required in type II systems. However, the PAM is recognized on the same strand as the crRNA (the opposite strand to type I systems).

Type III systems, like type I require six or seven Cas proteins binding to crRNAs. The type III systems analysed from *S. solfataricus* and *P. furiosus* both target the mRNA of phages rather than phage DNA genome, which may make these systems uniquely capable of targeting RNA-based phage genomes. Type III systems were also found to target DNA in addition to RNA using a different Cas protein in the complex, Cas10. The DNA cleavage was shown to be transcription dependent.

The mechanism for distinguishing self from foreign DNA during interference is built into the crRNAs and is therefore likely common to all three systems. Throughout the distinctive maturation process of each major type, all crRNAs contain a spacer sequence and some portion of the repeat at one or both ends. It is the partial repeat sequence that prevents the CRISPR-Cas system from targeting the chromosome as base pairing beyond the spacer sequence signals self and prevents DNA cleavage. RNA-guided CRISPR enzymes are classified as type V restriction enzymes.

6.3.4 Evolution

The cas genes in the adaptor and effector modules of the CRISPR-Cas system are believed to have evolved from two different ancestral modules. A transposon-like element called casposon encoding the Cas1-like integrase and potentially other components of the adaptation module was inserted next to the ancestral effector module, which likely functioned as an independent innate immune system. The highly conserved cas1 and cas2 genes of the adaptor module evolved from the ancestral module while a variety of class 1 effector cas genes evolved from the ancestral effector module. The evolution of these various class 1 effector module cas genes was guided by various mechanisms, such as duplication events. On the other hand, each type of class 2 effector module arose from subsequent independent insertions of mobile genetic elements. These mobile genetic elements took the place of the multiple gene effector modules to create single gene effector modules that produce large proteins which perform all the necessary tasks of the effector module. The spacer regions of CRISPR-Cas systems are taken directly from foreign mobile genetic elements and thus their long-term evolution is hard to trace. The non-random evolution of these spacer regions has been found to be highly dependent on the environment and the particular foreign mobile genetic elements it contains.

CRISPR-Cas can immunize bacteria against certain phages and thus halt transmission. For this reason, Koonin described CRISPR-Cas as a Lamarckian inheritance mechanism. However, this was disputed by a critic who noted, "We should remember [Lamarck] for the good he contributed to science, not for things that resemble his theory only superficially. Indeed, thinking of CRISPR and other phenomena as Lamarckian only obscures the simple and elegant way evolution really works". But as more recent studies have been conducted, it has become apparent that the acquired spacer regions of CRISPR-Cas systems are indeed a form of Lamarckian evolution because they are genetic mutations that are acquired and then passed on. On the other hand, the evolution of the Cas gene machinery that facilitates the system evolves through classic Darwinian evolution.

Coevolution

Analysis of CRISPR sequences revealed coevolution of host and viral genomes. The basic model of CRISPR evolution is newly incorporated spacers driving phages to mutate their genomes to avoid the bacterial immune response, creating diversity in both the phage and host populations. To resist a phage infection, the sequence of the CRISPR spacer must correspond perfectly to the sequence of the target phage gene. Phages can continue to infect their hosts' given point mutations in the spacer. Similar stringency is required in PAM or the bacterial strain remains phage sensitive.

Rates

A study of 124 *S. thermophilus* strains showed that 26% of all spacers were unique and that different CRISPR loci showed different rates of spacer acquisition. Some CRISPR loci evolve more rapidly than others, which allowed the strains' phylogenetic relationships to be determined. A comparative genomic analysis showed that *E. coli* and *S. enterica* evolve much more slowly than *S. thermophilus*. The latter's strains that diverged 250,000 years ago still contained the same spacer complement.

Metagenomic analysis of two acid-mine-drainage biofilms showed that one of the analyzed CRISPRs contained extensive deletions and spacer additions versus the other biofilm, suggesting a higher phage activity/prevalence in one community than the other. In the oral cavity, a temporal study determined that 7–22% of spacers were shared over 17 months within an individual while less than 2% were shared across individuals.

From the same environment, a single strain was tracked using PCR primers specific to its CRISPR system. Broad-level results of spacer presence/absence showed significant diversity. However, this CRISPR added three spacers over 17 months, suggesting that even in an environment with significant CRISPR diversity some loci evolve slowly.

CRISPRs were analysed from the metagenomes produced for the Human Microbiome Project. Although most were body-site specific, some within a body site are widely shared among individuals. One of these loci originated from streptococcal species and contained ≈15,000 spacers, 50% of which were unique. Similar to the targeted studies of the oral cavity, some showed little evolution over time.

CRISPR evolution was studied in chemostats using *S. thermophilus* to directly examine spacer acquisition rates. In one week, *S. thermophilus* strains acquired up to three spacers when challenged with a single phage. During the same interval, the phage developed single-nucleotide polymorphisms that became fixed in the population, suggesting that targeting had prevented phage replication absent these mutations.

Another *S. thermophilus* experiment showed that phages can infect and replicate in hosts that have only one targeting spacer. Yet another showed that sensitive hosts can exist in environments with high-phage titres. The chemostat and observational studies suggest many nuances to CRISPR and phage (co)evolution.

6.3.5 Identification

CRISPRs are widely distributed among bacteria and archaea and show some sequence similarities. Their most notable characteristic is their repeating spacers and direct repeats. This characteristic makes CRISPRs easily identifiable in long sequences of DNA, since the number of repeats decreases the likelihood of a false positive match.

Analysis of CRISPRs in metagenomic data is more challenging, as CRISPR loci do not typically assemble, due to their repetitive nature or through strain variation, which confuses assembly algorithms. Where many reference genomes are available, polymerase chain reaction (PCR) can be used to amplify CRISPR arrays and analyse spacer content. However, this approach yields information only for specifically targeted CRISPRs and for organisms with sufficient representation in public databases to design reliable polymerase PCR primers. Degenerate repeat-specific primers can be used to amplify CRISPR spacers directly from environmental samples; amplicons containing two or three spacers can be then computationally assembled to reconstruct long CRISPR arrays.

The alternative is to extract and reconstruct CRISPR arrays from shotgun metagenomic data. This is computationally more difficult, particularly with second generation sequencing technologies (e.g. 454, Illumina), as the short read lengths prevent more than two or three repeat units appearing in a single read. CRISPR identification in raw reads has been achieved using purely *de novo* identification or by using direct repeat sequences in partially assembled CRISPR arrays from contigs (overlapping DNA segments that together represent a consensus region of DNA) and direct repeat sequences from published genomes as a hook for identifying direct repeats in individual reads.

6.3.6 Use by Phages

Another way for bacteria to defend against phage infection is by having chromosomal islands. A subtype of chromosomal islands called phage-inducible chromosomal island (PICI) is excised from a bacterial chromosome upon phage infection and can inhibit phage replication. PICIs are induced, excised, replicated, and finally packaged into small capsids by certain staphylococcal temperate phages. PICIs use several mechanisms to block phage reproduction. In the first mechanism, PICI-encoded Ppi differentially blocks phage maturation by binding or interacting specifically with phage TerS, hence blocking phage TerS/TerL complex formation responsible for phage DNA packaging. In the second mechanism PICI CpmAB redirects the phage capsid morphogenetic protein to make 95% of SaPI-sized capsid and phage DNA can package only 1/3rd of their genome in these small capsids and hence become nonviable phage. The third mechanism involves two proteins, PtiA and PtiB, that target the LtrC, which is responsible for the production of virion and lysis proteins. This interference mechanism is modulated by a modulatory protein, PtiM, binds to one of the interference-mediating proteins, PtiA, and hence achieves the required level of interference.

One study showed that lytic ICP1 phage, which specifically targets *Vibrio cholerae* serogroup O1, has acquired a CRISPR-Cas system that targets a *V. cholera* PICI-like element. The system has 2 CRISPR loci and 9 Cas genes. It seems to be homologous to the I-F system found in *Yersinia pestis*. Moreover, like the bacterial CRISPR-Cas system, ICP1 CRISPR-Cas can acquire new sequences, which allows phage and host to co-evolve.

Certain archaeal viruses were shown to carry mini-CRISPR arrays containing one or two spacers. It has been shown that spacers within the virus-borne CRISPR arrays target other viruses and plasmids, suggesting that mini-CRISPR arrays represent a mechanism of heterotypic superinfection exclusion and participate in interviral conflicts.

6.3.7 Applications

CRISPR gene editing is a revolutionary technology that allows for precise, targeted modifications to the DNA of living organisms. Developed from a natural defense mechanism found in bacteria, CRISPR-Cas9 is the most commonly used system, that allows "cutting" of DNA at specific locations and either delete, modify, or insert genetic material. This technology has transformed fields such as genetics, medicine, and agriculture, offering potential treatments for genetic disorders, advancements in crop engineering, and research into the fundamental workings of life. However, its ethical implications and potential unintended consequences have sparked significant debate.

6.4 TALENS AND ZFNS: PRECISION GENE EDITING PROTEINS

Before CRISPR, two gene-editing proteins, TALENs (Transcription Activator-Like Effector Nucleases) and ZFNs (Zinc Finger Nucleases), were used to create targeted DNA modifications. Both TALENs and ZFNs are engineered proteins that bind to specific DNA sequences and induce double-strand breaks. Each TALEN or ZFN molecule can be customized to recognize unique DNA sequences, offering a high degree of control over gene editing.

While TALENs and ZFNs represented significant advancements in gene editing, they required complex protein engineering, which made them challenging and time-consuming to design. Although they are still used in specific applications where CRISPR may not be as effective, TALENs and ZFNs have generally been overshadowed by the simplicity and versatility of CRISPR-Cas9. However, they remain valuable tools for scientists working in specialized areas of recombinant DNA technology, particularly when working with genomes where CRISPR is less efficient or precise.

6.4.1 Transcription Activator-Like Effector Nuclease

Transcription activator-like effector nucleases (TALEN) are restriction enzymes that can be engineered to cut specific sequences of DNA. They are made by fusing a TAL effector DNA-binding domain to a DNA cleavage domain (a nuclease which cuts DNA strands). Transcription activator-like effectors (TALEs) can be engineered to bind to practically any desired DNA sequence, so when combined with a nuclease, DNA can be cut at specific locations. The restriction enzymes can be introduced into cells, for use in gene editing or for genome editing *in situ*, a technique known as genome editing with engineered nucleases. Alongside zinc finger nucleases and CRISPR/Cas9, TALEN is a prominent tool in the field of genome editing.

TALE DNA-Binding Domain

TAL effectors are proteins that are secreted by *Xanthomonas* bacteria via their type III secretion system when they infect plants. The DNA binding domain contains a repeated highly conserved 33–34 amino acid sequence with divergent 12th and 13th amino acids. These two positions, referred to as the Repeat Variable Diresidue

(RVD), are highly variable and show a strong correlation with specific nucleotide recognition. This straightforward relationship between amino acid sequence and DNA recognition has allowed for the engineering of specific DNA-binding domains by selecting a combination of repeat segments containing the appropriate RVDs. Notably, slight changes in the RVD and the incorporation of "nonconventional" RVD sequences can improve targeting specificity.

DNA Cleavage Domain

The non-specific DNA cleavage domain from the end of the FokI endonuclease can be used to construct hybrid nucleases that are active in a yeast assay. These reagents are also active in plant cells and in animal cells. Initial TALEN studies used the wild-type FokI cleavage domain, but some subsequent TALEN studies also used FokI cleavage domain variants with mutations designed to improve cleavage specificity and cleavage activity. The FokI domain functions as a dimer, requiring two constructs with unique DNA binding domains for sites in the target genome with proper orientation and spacing. Both the number of amino acid residues between the TALE DNA binding domain and the FokI cleavage domain and the number of bases between the two individual TALEN binding sites appear to be important parameters for achieving high levels of activity.

Engineering TALEN Constructs

The simple relationship between amino acid sequence and DNA recognition of the TALE binding domain allows for the efficient engineering of proteins. In this case, artificial gene synthesis is problematic because of improper annealing of the repetitive sequence found in the TALE binding domain. One solution to this is to use a publicly available software program (DNAWorks) to calculate oligonucleotides suitable for assembly in a two step PCR oligonucleotide assembly followed by whole gene amplification. A number of modular assembly schemes for generating engineered TALE constructs have also been reported. Both methods offer a systematic approach to engineering DNA binding domains that is conceptually similar to the modular assembly method for generating zinc finger DNA recognition domains.

Transfection

Once the TALEN constructs have been assembled, they are inserted into plasmids; the target cells are then transfected with the plasmids, and the gene products are expressed and enter the nucleus to access the genome. Alternatively, TALEN constructs can be delivered to the cells as mRNAs, which removes the possibility of genomic integration of the TALEN-expressing protein. Using an mRNA vector can also dramatically increase the level of homology directed repair (HDR) and the success of introgression during gene editing.

Genome Editing

Mechanisms

TALEN can be used to edit genomes by inducing double-strand breaks (DSB), which cells respond to with repair mechanisms. Non-homologous end joining (NHEJ) directly ligates DNA from either side of a double-strand break where there is very little or no sequence overlap for annealing.

This repair mechanism induces errors in the genome via indels (insertion or deletion), or chromosomal rearrangement; any such errors may render the gene products coded at that location non-functional. Because this activity can vary depending on the species, cell type, target gene, and nuclease used, it should be monitored when designing new systems. A simple heteroduplex cleavage assay can be run which detects any difference between two alleles amplified by PCR. Cleavage products can be visualized on simple agarose gels or slab gel systems. Alternatively, DNA can be introduced into a genome through NHEJ in the presence of exogenous double-stranded DNA fragments. Homology directed repair can also introduce foreign DNA at the DSB as the transfected double-stranded sequences are used as templates for the repair enzymes.

Applications

TALEN has been used to efficiently modify plant genomes, creating economically important food crops with favorable nutritional qualities. They have also been harnessed to develop tools for the production of biofuels. In addition, it has been used to engineer stably modified human embryonic stem cell and induced pluripotent stem cell (IPSCs) clones and human erythroid cell lines, to generate knockout *C. elegans*, knockout rats, knockout mice, and knockout zebrafish. Moreover, the method can be used to generate knockin organisms. Wu et al.obtained a Sp110 knockin cattle using Talen nickases to induce increased resistance of tuberculosis. This approach has also been used to generate knockin rats by TALEN mRNA microinjection in one-cell embryos.

TALEN has also been utilized experimentally to correct the genetic errors that underlie disease. For example, it has been used *in vitro* to correct the genetic defects that cause disorders such as sickle cell disease, xeroderma pigmentosum, and epidermolysis bullosa. Recently, it was shown that TALEN can be used as tools to harness the immune system to fight cancers; TALEN-mediated targeting can generate T cells that are resistant to chemotherapeutic drugs and show anti-tumor activity.

In theory, the genome-wide specificity of engineered TALEN fusions allows for correction of errors at individual genetic loci via homology-directed repair from a correct exogenous template. In reality, however, the *in situ* application of TALEN is currently limited by the lack of an efficient delivery mechanism, unknown immunogenic factors, and uncertainty in the specificity of TALEN binding.

Another emerging application of TALEN is its ability to combine with other genome engineering tools, such as meganucleases. The DNA binding region of a TAL effector can be combined with the cleavage domain of a meganuclease to create a hybrid architecture combining the ease of engineering and highly specific DNA binding activity of a TAL effector with the low site frequency and specificity of a meganuclease.

In comparison to other genome editing techniques, TALEN falls in the middle in terms of difficulty and cost. Unlike ZFNs, TALEN recognizes single nucleotides. It's far more straightforward to engineer interactions between TALEN DNA binding domains and their target nucleotides than it is to create interactions with ZFNs and their target nucleotide triplets. On the other hand, CRISPR relies on ribonucleotide complex formation instead of protein/DNA recognition. gRNAs have occasionally limitations regarding feasibility due to lack of PAM sites in the target sequence and even though they can be cheaply produced, the current development lead to a remarkable decrease of cost for TALENs, so that they are in a similar price and time range like CRISPR based genome editing.

TAL Effector Nuclease Precision

The off-target activity of an active nuclease may lead to unwanted double-strand breaks and may consequently yield chromosomal rearrangements and/or cell death. Studies have been carried out to compare the relative nuclease-associated toxicity of available technologies. Based on these studies and the maximal theoretical distance between DNA binding and nuclease activity, TALEN constructs are believed to have the greatest precision of the currently available technologies.

6.4.2 Zinc-Finger Nuclease

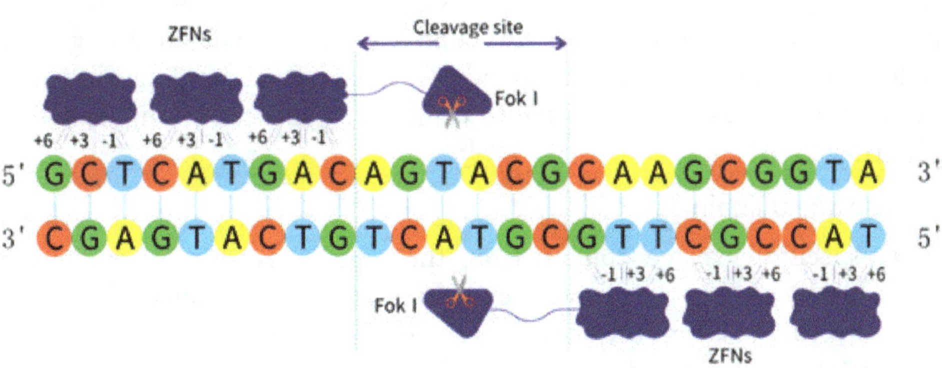

Fig: The Schematic Diagram of ZFNs

Zinc-finger nucleases (ZFNs) are artificial restriction enzymes generated by fusing a zinc finger DNA-binding domain to a DNA-cleavage domain. Zinc finger domains can be engineered to target specific desired DNA sequences and this enables zinc-finger nucleases to target unique sequences within complex genomes. By taking advantage of endogenous DNA repair machinery, these reagents can be used to precisely alter the genomes of higher organisms. Alongside CRISPR/Cas9 and TALEN, ZFN is a prominent tool in the field of genome editing. It was initially created by researcher Srinivasan Chandrasegaran.

Domains

DNA-Binding Domain

The DNA-binding domains of individual ZFNs typically contain between three and six individual zinc finger repeats and can each recognize between 9 and 18 basepairs. If the zinc finger domains perfectly recognize a 3 basepair DNA sequence, they can generate a 3-finger array that can recognize a 9 basepair target site. Other procedures can utilize either 1-finger or 2-finger modules to generate zinc-finger arrays with six or more individual zinc fingers. The main drawback with this procedure is the specificities of individual zinc fingers can overlap and can depend on the context of the surrounding zinc fingers and DNA. Without methods to account for this "context dependence", the standard modular assembly procedure often fails.

Numerous selection methods have been used to generate zinc-finger arrays capable of targeting desired sequences. Initial selection efforts utilized phage display to select proteins that bound a given DNA target from a large pool of partially randomized zinc-finger arrays. More recent efforts have utilized yeast one-hybrid systems, bacterial one-hybrid and two-hybrid systems, and mammalian cells. A promising new method to select novel zinc-finger arrays utilizes a bacterial two-hybrid system and has been dubbed "OPEN" by its creators. This system combines pre-selected pools of individual zinc fingers that were each selected to bind a given triplet and then utilizes a second round of selection to obtain 3-finger arrays capable of binding a desired 9-bp sequence. This system was developed by the Zinc-Finger Consortium as an alternative to commercial sources of engineered zinc-finger arrays.

DNA-Cleavage Domain

The non-specific cleavage domain from the type IIs restriction endonuclease FokI is typically used as the cleavage domain in ZFNs. This cleavage domain must dimerize in order to cleave DNA and thus a pair of ZFNs are required to target non-palindromic DNA sites. Standard ZFNs fuse the cleavage domain to the C-terminus of each zinc finger domain. To let the two cleavage domains dimerize and cleave DNA, the two individual ZFNs must bind opposite strands of DNA with their C-termini a certain distance apart. The most commonly used linker sequences between the zinc finger domain and the cleavage domain requires the 5' edge of each binding site to be separated by 5 to 7 bp.

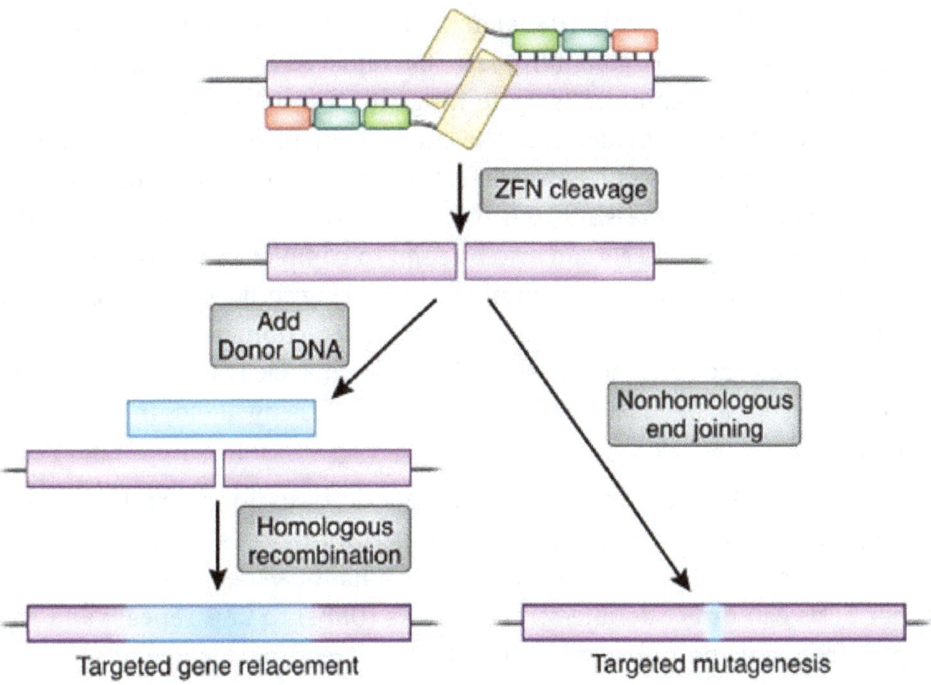

Fig: A Pair of ZFNs, Each with Three Zinc Fingers Binding to Target DNA, are shown Introducing a Double-Strand Break, at the FokI Domain, Depicted in Yellow. Subsequently, the Double Strand Break is Shown as being Repaired through Either Homology-Directed Repair or Non-Homologous end Joining

Several different protein engineering techniques have been employed to improve both the activity and specificity of the nuclease domain used in ZFNs. Directed evolution has been employed to generate a FokI variant with enhanced cleavage activity that the authors dubbed "Sharkey". Structure-based design has also been employed to improve the cleavage specificity of FokI by modifying the dimerization interface so that only the intended heterodimeric species are active.

Applications

Zinc finger nucleases are useful to manipulate the genomes of many plants and animals including arabidopsis, tobacco, soybean, corn, *Drosophila melanogaster*, *C. elegans*, *Platynereis dumerilii*, sea urchin, silkworm, zebrafish, frogs, mice, rats, rabbits, pigs, cattle, and various types of mammalian cells. Zinc finger nucleases have also been used in a mouse model of haemophilia and a clinical trial found CD4+ human T-cells with the CCR5 gene disrupted by zinc finger nucleases to be safe as a potential treatment for HIV/AIDS. ZFNs are also used to create a new generation of genetic disease models called isogenic human disease models.

Disabling an Allele

ZFNs can be used to disable dominant mutations in heterozygous individuals by producing double-strand breaks (DSBs) in the DNA in the mutant allele, which will, in the absence of a homologous template, be repaired by non-homologous end-joining (NHEJ). NHEJ repairs DSBs by joining the two ends together and usually produces no mutations, provided that the cut is clean and uncomplicated. In some instances, however, the repair is imperfect, resulting in deletion or insertion of base-pairs, producing frame-shift and preventing the production of the harmful protein. Multiple pairs of ZFNs can also be used to completely remove entire large segments of genomic sequence. To monitor the editing activity, a PCR of the target area amplifies both alleles and, if one contains an insertion, deletion, or mutation, it results in a heteroduplex single-strand bubble that cleavage assays can easily detect. ZFNs have also been used to modify disease-causing alleles in triplet repeat disorders.

Expanded CAG/CTG repeat tracts are the genetic basis for more than a dozen inherited neurological disorders including Huntington's disease, myotonic dystrophy, and several spinocerebellar ataxias. It has been demonstrated in human cells that ZFNs can direct double-strand breaks (DSBs) to CAG repeats and shrink the repeat from long pathological lengths to short, less toxic lengths.

Recently, a group of researchers have successfully applied the ZFN technology to genetically modify the gol pigment gene and the ntl gene in zebrafish embryo. Specific zinc-finger motifs were engineered to recognize distinct DNA sequences. The ZFN-encoding mRNA was injected into one-cell embryos and a high percentage of animals carried the desired mutations and phenotypes. Their research work demonstrated that ZFNs can specifically and efficiently create heritable mutant alleles at loci of interest in the germ line, and ZFN-induced alleles can be propagated in subsequent generations.

Similar research of using ZFNs to create specific mutations in zebrafish embryo has also been carried out by other research groups. The kdr gene in zebra fish encodes for the vascular endothelial growth factor-2 receptor. Mutagenic lesions at this target site was induced using ZFN technique by a group of researchers in US. They suggested that the ZFN technique allows straightforward generation of a targeted allelic series of mutants; it does not rely on the existence of species-specific embryonic stem cell lines and is applicable to other vertebrates, especially those whose embryos are easily available; finally, it is also feasible to achieve targeted knock-ins in zebrafish, therefore it is possible to create human disease models that are heretofore inaccessible.

Allele Editing

ZFNs are also used to rewrite the sequence of an allele by invoking the homologous recombination (HR) machinery to repair the DSB using the supplied DNA fragment as a template. The HR machinery searches for homology between the damaged chromosome and the extra-chromosomal fragment and copies the sequence of the fragment between the two broken ends of the chromosome, regardless of whether the fragment contains the original sequence. If the subject is homozygous for the target allele, the efficiency of the technique is reduced since the undamaged copy of the allele may be used as a template for repair instead of the supplied fragment.

Gene Therapy

The success of gene therapy depends on the efficient insertion of therapeutic genes at an appropriate chromosomal target site within the human genome, without causing cell injury, oncogenic mutations, or an immune response. The construction of plasmid vectors is simple and straightforward. Custom-designed ZFNs that combine the non-specific cleavage domain (N) of *Fok*I endonuclease with zinc-finger proteins (ZFPs) offer a general way to deliver a site-specific DSB to the genome, and stimulate local homologous recombination by several orders of magnitude. This makes targeted gene correction or genome editing a viable option in human cells. Since ZFN-encoding plasmids could be used to transiently express ZFNs to target a DSB to a specific gene locus in human cells, they offer an excellent way for targeted delivery of the therapeutic genes to a pre-selected chromosomal site. The ZFN-encoding plasmid-based approach has the potential to circumvent all the problems associated with the viral delivery of therapeutic genes. The first therapeutic applications of ZFNs are likely to involve *ex vivo* therapy using a patient's own stem cells. After editing the stem cell genome, the cells could be expanded in culture and reinserted into the patient to produce differentiated cells with corrected functions. Initial targets likely include the causes of monogenic diseases, such as the IL2Rγ gene and the β-globin gene for gene correction and CCR5 gene for mutagenesis and disablement.

Potential Problems

Off-Target Cleavage

If the zinc finger domains are not specific enough for their target site or they do not target a unique site within the genome of interest, off-target cleavage may occur. Such off-target cleavage may lead to the production of enough double-strand breaks to overwhelm the repair machinery and, as a consequence, yield chromosomal rearrangements and/or cell death. Off-target cleavage events may also promote random integration of donor DNA. Two separate methods have been demonstrated to decrease off-target cleavage for 3-finger ZFNs that target two adjacent 9-basepair sites. Other groups use ZFNs with 4, 5 or 6 zinc fingers that target longer and presumably rarer sites and such ZFNs could theoretically yield less off-target activity. A comparison of a pair of 3-finger ZFNs and a pair of 4-finger ZFNs detected off-target cleavage in human cells at 31 loci for the 3-finger ZFNs and at 9 loci for the 4-finger ZFNs. Whole genome sequencing of *C. elegans* modified with a pair of 5-finger ZFNs found only the intended modification and a deletion at a site "unrelated to the ZFN site" indicating this pair of ZFNs was capable of targeting a unique site in the *C. elegans* genome.

Immunogenicity

As with many foreign proteins inserted into the human body, there is a risk of an immunological response against the therapeutic agent and the cells in which it is active. Since the protein must be expressed only transiently, however, the time over which a response may develop is short. Liu et al. respectively target ZFNickases to the endogenous b-casein(CSN2) locus stimulates lysostaphin and human lysozyme gene addition by homology-directed repair and derive secrete lysostaphin cows.

Prospects

The ability to precisely manipulate the genomes of plants and animals has numerous applications in basic research, agriculture, and human therapeutics. Using ZFNs to modify endogenous genes has traditionally been a difficult task due mainly to the challenge of generating zinc finger domains that target the desired sequence with sufficient specificity. Improved methods of engineering zinc finger domains and the availability of ZFNs from a commercial supplier now put this technology in the hands of increasing numbers of researchers. Several groups are also developing other types of engineered nucleases including engineered homing endonucleases and nucleases based on engineered TAL effectors. TAL effector nucleases (TALENs) are particularly interesting because TAL effectors appear to be very simple to engineer and TALENs can be used to target endogenous loci in human cells. But to date no one has reported the isolation of clonal cell lines or transgenic organisms using such reagents. One type of ZFN, known as SB-728-T, has been tested for potential application in the treatment of HIV.

Zinc-Finger Nickases

Zinc-finger nickases (ZFNickases) are created by inactivating the catalytic activity of one ZFN monomer in the ZFN dimer required for double-strand cleavage. ZFNickases demonstrate strand-specific nicking activity *in vitro* and thus provide for highly specific single-strand breaks in DNA. These SSBs undergo the same cellular mechanisms for DNA that ZFNs exploit, but they show a significantly reduced frequency of mutagenic NHEJ repairs at their target nicking site. This reduction provides a bias for HR-mediated gene modifications. ZFNickases can induce targeted HR in cultured human and livestock cells, although at lower levels than corresponding ZFNs from which they were derived because nicks can be repaired without genetic alteration. A major limitation of ZFN-mediated gene modifications is the competition between NHEJ and HR repair pathways. Regardless of the presence of a DNA donor construct, both repair mechanisms can be activated following DSBs induced by ZFNs.

Thus, ZFNickases is the first plausible attempt at engineering a method to favor the HR method of DNA repair as opposed to the error-prone NHEJ repair. By reducing NHEJ repairs, ZFNickases can thereby reduce the spectrum of unwanted off-target alterations. The ease by which ZFNickases can be derive from ZFNs provides a great platform for further studies regarding the optimization of ZFNickases and possibly increasing their levels of targeted HR while still maintain their reduced NHEJ frequency.

6.5 RECOMBINASE ENZYMES: A TOOL FOR DNA REARRANGEMENTS

Recombinase enzymes, such as Cre and FLP, are proteins that catalyze site-specific recombination, allowing for precise rearrangement of DNA within an organism's genome. They operate by recognizing specific DNA sequences called loxP or FRT sites, where they facilitate the excision, inversion, or integration of DNA fragments. This technology is widely used in genetic engineering to create conditional knockout models, where a gene can be "turned off" or modified at specific developmental stages or in particular tissues.

Recombinase enzymes add flexibility to the recombinant DNA toolbox by allowing for complex gene rearrangements that are otherwise challenging to achieve. They are particularly useful in developmental biology and neuroscience, where scientists can study gene function in a spatially and temporally controlled manner. While recombinase technology has limitations, such as requiring the presence of target sites in the genome, it remains a valuable tool for researchers working with genetically engineered animal models.

6.6 SYNTHETIC BIOLOGY: DE NOVO DNA SYNTHESIS

Advancements in synthetic biology have introduced the possibility of de novo DNA synthesis, allowing scientists to design and assemble entirely new DNA sequences from scratch. This goes beyond simply editing existing genes to creating new genetic sequences tailored for specific functions. Using automated synthesis machines, researchers can construct DNA sequences that code for novel proteins, metabolic pathways, or regulatory circuits. This technology has applications in bioengineering, where custom-made genes can be used to produce biofuels, pharmaceuticals, or other valuable compounds.

De novo DNA synthesis also enables the creation of synthetic organisms, designed with minimal genomes containing only essential genes. These organisms can serve as biological "factories," producing useful compounds or studying basic life processes. While de novo synthesis is still relatively costly, advances in the field continue to drive down costs and increase accessibility. This capability to design genes from the ground up represents an exciting frontier in recombinant DNA technology, with broad implications for biotechnology and medicine.

6.7 VIRAL VECTORS: EFFICIENT GENE DELIVERY SYSTEMS

Viral vectors, derived from viruses such as lentivirus, adenovirus, and adeno-associated virus (AAV), are powerful tools for delivering genetic material into host cells. Viruses naturally evolved to deliver their genetic material into host cells, making them effective vehicles for gene delivery in genetic therapy and research. By modifying these viruses to carry specific genes, scientists can use them to introduce recombinant DNA into cells with high efficiency.

Each viral vector has unique characteristics, making it suitable for different applications. For instance, lentiviruses can integrate into the host genome, providing long-term expression, while AAVs are often used in gene therapy due to their low immunogenicity. Viral vectors are essential in gene therapy research, where they are used to deliver corrective genes to patients with genetic disorders. Their addition to the recombinant DNA toolbox provides a reliable method for achieving stable gene expression in various cell types.

6.8 RNA INTERFERENCE (RNAI) AND CRISPR INTERFERENCE (CRISPRI)

RNA interference (RNAi) and CRISPR interference (CRISPRi) are technologies that allow scientists to "turn off" genes temporarily rather than deleting them permanently. RNAi uses small RNA molecules to target specific mRNA transcripts for degradation, preventing them from being translated into proteins. CRISPRi, a variation of CRISPR technology, uses a modified Cas9 protein that binds to DNA but does not cut it, blocking transcription without altering the DNA sequence.

These tools are invaluable for studying gene function, as they enable researchers to observe the effects of reduced gene expression without permanently modifying the genome. RNAi and CRISPRi are widely used in functional genomics, disease modeling, and drug discovery, providing a level of control that complements traditional gene knockout methods. This flexibility makes RNAi and CRISPRi essential components of the modern recombinant DNA toolkit.

6.9 THE FUTURE OF RECOMBINANT DNA TECHNOLOGY

The field of recombinant DNA technology continues to evolve, with new tools emerging that promise even greater control over genetic material. Innovations such as base editing, prime editing, and CRISPR-associated transposases offer more refined methods for DNA manipulation, allowing scientists to edit single nucleotides or insert genes with unprecedented accuracy. These tools hold great potential for medical applications, as they could enable precise correction of genetic mutations in patients with inherited disorders.

Additionally, the integration of recombinant DNA with bioinformatics and machine learning is leading to new approaches for designing synthetic genes and predicting the effects of genetic modifications. As these technologies advance, the recombinant DNA toolbox will continue to expand, offering even more powerful methods for studying and manipulating genetic material.

6.10 CONCLUSION

The recombinant DNA toolbox has undergone significant expansion over the past few decades, with innovative tools such as CRISPR-Cas9, TALENs, recombinase enzymes, synthetic DNA synthesis, viral vectors, and RNA interference techniques transforming the field of genetic engineering. These tools have made it possible to manipulate DNA with greater precision, efficiency, and flexibility, enabling breakthroughs in medicine, agriculture, and biotechnology. As researchers continue to develop and refine these tools, the potential for recombinant DNA technology to address pressing global challenges—such as genetic diseases, food security, and sustainable energy production—will only grow. The future of recombinant DNA is bright, promising new insights into genetics and unprecedented opportunities for improving human health and welfare.

CHAPTER 7

MOBILE DNA SEQUENCES IN THE GENOME

7.1 INTRODUCTION TO MOBILE DNA ELEMENTS

Mobile DNA sequences, often referred to as transposable elements (TEs) or "jumping genes," are DNA sequences capable of moving or copying themselves within a genome. These sequences were first discovered by geneticist Barbara McClintock in the 1940s through her work on maize, earning her a Nobel Prize in 1983. The discovery of mobile DNA sequences has transformed our understanding of genome structure, dynamics, and evolution. TEs play significant roles in the diversity and adaptability of organisms, influencing gene expression, mutations, and the overall architecture of genomes. Despite their contributions, TEs can sometimes disrupt normal cellular functions, leading to various genomic instabilities and diseases.

7.2 TYPES OF MOBILE DNA SEQUENCES

Mobile DNA sequences are broadly categorized into two main classes based on their transposition mechanisms: Class I elements, or retrotransposons, and Class II elements, or DNA transposons.

7.2.1 Class I: Retrotransposons

Retrotransposons, which make up a large portion of mobile DNA in eukaryotic genomes, function via a "copy-and-paste" mechanism. These elements are transcribed into RNA, which is then reverse-transcribed back into DNA and inserted at a new location within the genome.

Long Terminal Repeat (LTR) Retrotransposons

A long terminal repeat (LTR) is a pair of identical sequences of DNA, several hundred base pairs long, which occur in eukaryotic genomes on either end of a series of genes or pseudogenes that form a retrotransposon or an endogenous retrovirus or a retroviral provirus. All retroviral genomes are flanked by LTRs, while there are some retrotransposons without LTRs. Typically, an element flanked by a pair of LTRs will encode a reverse transcriptase and an integrase, allowing the element to be copied and inserted at a different location of the genome. Copies of such an LTR-flanked element can often be found hundreds or thousands of times in a genome. LTR retrotransposons comprise about 8% of the human genome. The first LTR sequences were found by A.P. Czernilofsky and J. Shine in 1977 and 1980.

Transcription

The LTR-flanked sequences are partially transcribed into an RNA intermediate, followed by reverse transcription into complementary DNA (cDNA) and ultimately dsDNA (double-stranded DNA) with full LTRs. The LTRs then mediate integration of the DNA via an LTR specific integrase into another region of the host chromosome. Retroviruses such as human immunodeficiency virus (HIV) use this basic mechanism.

Dating Retroviral Insertions

As 5' and 3' LTRs are identical upon insertion, the difference between paired LTRs can be used to estimate the age of ancient retroviral insertions. This method of dating is used by paleovirologists, though it fails to take into account confounding factors such as gene conversion and homologous recombination.

HIV-1

The HIV-1 LTR is 634 bp in length and, like other retroviral LTRs, is segmented into the U3, R, and U5 regions. U3 and U5 has been further subdivided according to transcription factor sites and their impact on LTR activity and viral gene expression. The multi-step process of reverse transcription results in the placement of two identical LTRs, each consisting of a U3, R, and U5 region, at either end of the proviral DNA. The ends of the LTRs subsequently participate in integration of the provirus into the host genome. Once the provirus has been integrated, the LTR on the 5' end serves as the promoter for the entire retroviral genome, while the LTR at the 3' end provides for nascent viral RNA polyadenylation and, in HIV-1, HIV-2, and SIV, encodes the accessory protein, Nef.

All of the required signals for gene expression are found in the LTRs: Enhancer, promoter (can have both transcriptional enhancers or regulatory elements), transcription initiation (such as capping), transcription terminator and polyadenylation signal.

In HIV-1, the 5'UTR region has been characterized according to functional and structural differences into several sub-regions:

- TAR, or trans-activation response element, plays a critical role in transcriptional activation via its interaction with viral proteins. It forms a highly stable stem–loop structure consisting of 26 base pairs with a bulge in its secondary structure that interfaces with the viral transcription activator protein Tat.
- Poly a plays roles both in dimerization and genome packaging since

it is necessary for cleavage and polyadenylation. It has been reported that sequences upstream (U3 region) and downstream (U5 region) are needed in order to make the cleavage process efficient.

- PBS, or primer binding site, is 18 nucleotides long and has a specific sequence that binds to the $tRNA^{Lys}$ primer required for initiation of reverse transcription.
- Psi (Ψ), or the Psi packaging element, is a unique motif involved in regulating the packaging of the viral genome into the capsid. It is composed of four stem-loop (SL) structures with a major splicing donor site embedded in the second SL.
- DIS, or dimer initiation site, is a highly conserved RNA–RNA interacting sequence constituting the SL1 stem–loop in the Psi packaging element of many retroviruses. DIS is characterized by a conserved stem and palindromic loop that forms a kissing-loop complex between HIV-1 RNA genomes to dimerize them for encapsidation.

The transcript begins, at the beginning of R, is capped, and proceeds through U5 and the rest of the provirus, usually terminating by the addition of a poly A tract just after the R sequence in the 3' LTR.

The finding that both HIV LTRs can function as transcriptional promoters is not surprising since both elements are apparently identical in nucleotide sequence. Instead, the 3' LTR acts in transcription termination and polyadenylation. However, it has been suggested that the transcriptional activity of the 5' LTR is far greater than that of the 3' LTR, a situation that is very similar to that of other retroviruses.

During transcription of the human immunodeficiency virus type 1 provirus, polyadenylation signals present in the 5' long terminal repeat (LTR) are disregarded while the identical polyadenylation signals present in the 3'LTR are utilized efficiently. It has been suggested that transcribed sequences present within the HIV-1 LTR U3 region act in cis to enhance polyadenylation within the 3' LTR.

Non-LTR Retrotransposons

Retrotransposons (also called Class I transposable elements) are mobile elements which move in the host genome by converting their transcribed RNA into DNA through the reverse transcription. Thus, they differ from Class II transposable elements, or DNA transposons, in utilizing an RNA intermediate for the transposition and leaving the transposition donor site unchanged.

Through reverse transcription, retrotransposons amplify themselves quickly to become abundant in eukaryotic genomes such as maize (49–78%) and humans (42%). They are only present in eukaryotes but share features with retroviruses such as HIV, for example, discontinuous reverse transcriptase-mediated extrachromosomal recombination.

There are two main types of retrotransposons, long terminal repeats (LTRs) and non-long terminal repeats (non-LTRs). Retrotransposons are classified based on sequence and method of transposition. Most retrotransposons in the maize genome are LTR, whereas in humans they are mostly non-LTR.

LTR Retrotransposons

LTR retrotransposons are characterized by their long terminal repeats (LTRs), which are present at both the 5' and 3' ends of their sequences. These LTRs contain the promoters for these transposable elements (TEs), are essential for TE integration, and can vary in length from just over 100 base pairs (bp) to more than 1,000 bp. On average, LTR retrotransposons span several thousand base pairs, with the largest known examples reaching up to 30 kilobases (kb).

LTRs are highly functional sequences, and for that reason LTR and non-LTR retrotransposons differ greatly in their reverse transcription and integration mechanisms. Non-LTR retrotransposons use a target-primed reverse transcription (TPRT) process, which requires the RNA of the TE to be brought to the cleavage site of the retrotransposon's integrase, where it is reverse transcribed. In contrast, LTR retrotransposons undergo reverse transcription in the cytoplasm, utilizing two rounds of template switching, and a formation of a pre-integration complex (PIC) composed of double-stranded DNA and an integrase dimer bound to LTRs. This complex then moves into the nucleus for integration into a new genomic location.

LTR retrotransposons typically encode the proteins gag and pol, which may be combined into a single open reading frame (ORF) or separated into distinct ORFs. Similar to retroviruses, the gag protein is essential for capsid assembly and the packaging of the TE's RNA and associated proteins. The pol protein is necessary for reverse transcription and includes these crucial domains: PR (protease), RT (reverse transcriptase), RH (RNase H), and INT (integrase). Additionally, some LTR retrotransposons have an ORF for an envelope (env) protein that is incorporated into the assembled capsid, facilitating attachment to cellular surfaces.

Endogenous Retrovirus

An endogenous retrovirus is a retrovirus without virus pathogenic effects that has been integrated into the host genome by inserting their inheritable genetic information into cells that can be passed onto the next generation like a retrotransposon. Because of this, they share features with retroviruses and retrotransposons. When the retroviral DNA is integrated into the host genome they evolve into endogenous retroviruses that influence eukaryotic genomes. So many endogenous retroviruses have inserted themselves into eukaryotic genomes that they allow insight into biology between viral-host interactions and the role of retrotransposons in evolution and disease. Many retrotransposons share features with endogenous retroviruses, the property of recognising and fusing with the host genome.

However, there is a key difference between retroviruses and retrotransposons, which is indicated by the env gene. Although similar to the gene carrying out the same function in retroviruses, the env gene is used to determine whether the gene is retroviral or retrotransposon. If the gene is retroviral it can evolve from a retrotransposon into a retrovirus. They differ by the order of sequences in pol genes. Env genes are found in LTR retrotransposon types Ty1-copia (Pseudoviridae), Ty3-gypsy (Metaviridae) and BEL/Pao. They encode glycoproteins on the retrovirus envelope needed for entry into the host cell.

Retroviruses can move between cells whereas LTR retrotransposons can only move themselves into the genome of the same cell. Many vertebrate genes were formed from retroviruses and LTR retrotransposons. One endogenous retrovirus or LTR retrotransposon has the same function and genomic locations in different species, suggesting their role in evolution.

Non-LTR Retrotransposons

Like LTR retrotransposons, non-LTR retrotransposons contain genes for reverse transcriptase, RNA-binding protein, nuclease, and sometimes ribonuclease H domain but they lack the long terminal repeats.

RNA-binding proteins bind the RNA-transposition intermediate and nucleases are enzymes that break phosphodiester bonds between nucleotides in nucleic acids. Instead of LTRs, non-LTR retrotransposons have short repeats that can have an inverted order of bases next to each other aside from direct repeats found in LTR retrotransposons that is just one sequence of bases repeating itself.

Although they are retrotransposons, they cannot carry out reverse transcription using an RNA transposition intermediate in the same way as LTR retrotransposons. Those two key components of the retrotransposon are still necessary but the way they are incorporated into the chemical reactions is different. This is because unlike LTR retrotransposons, non-LTR retrotransposons do not contain sequences that bind tRNA. They mostly fall into two types – LINEs (Long interspersed nuclear elements) and SINEs (Short interspersed nuclear elements). SVA elements are the exception between the two as they share similarities with both LINEs and SINEs, containing Alu elements and different numbers of the same repeat. SVAs are shorter than LINEs but longer than SINEs. While historically viewed as "junk DNA", research suggests in some cases, both LINEs and SINEs were incorporated into novel genes to form new functions.

- **LINEs:**
  - When a LINE is transcribed, the transcript contains an RNA polymerase II promoter that ensures LINEs can be copied into whichever location it inserts itself into. RNA polymerase II is the enzyme that transcribes genes into mRNA transcripts. The ends of LINE transcripts are rich in multiple adenines, the bases that are added at the end of transcription so that LINE transcripts would not be degraded. This transcript is the RNA transposition intermediate.
  - The RNA transposition intermediate moves from the nucleus into the cytoplasm for translation. This gives the two coding regions of a LINE that in turn binds back to the RNA it is transcribed from. The LINE RNA then moves back into the nucleus to insert into the eukaryotic genome.
  - LINEs insert themselves into regions of the eukaryotic genome that are rich in bases AT. At AT regions LINE uses its nuclease to cut one strand of the eukaryotic double-stranded DNA. The adenine-rich sequence in LINE transcript base pairs with the cut strand to flag where the LINE will be inserted with hydroxyl groups. Reverse transcriptase recognises these hydroxyl groups to synthesise LINE retrotransposon where the DNA is cut. Like with LTR retrotransposons, this new inserted LINE contains eukaryotic genome information so it can be copied and pasted into other genomic regions easily. The information sequences are longer and more variable than those in LTR retrotransposons.
  - Most LINE copies have variable length at the start because reverse transcription usually stops before DNA synthesis is complete. In some cases this causes RNA polymerase II promoter to be lost so LINEs cannot transpose further.
- **Human L1:**
  - LINE-1 (L1) retrotransposons make up a significant portion of the human genome, with an estimated 500,000 copies per genome. Genes encoding for human LINE1 usually have their transcription inhibited by methyl groups binding to its DNA carried out by PIWI proteins and enzymes DNA methyltransferases. L1 retrotransposition can disrupt the nature of genes transcribed by pasting themselves inside or near genes which could in turn lead to human disease. LINE1s can only retrotranspose in some cases to form different chromosome structures contributing to differences in genetics between individuals. There is an estimate of 80–100 active L1s in the reference genome of the Human Genome Project, and an even smaller number of L1s within those active L1s retrotranspose often. L1 insertions have been associated with tumorigenesis by activating cancer-related genes oncogenes and diminishing tumor suppressor genes.
  - Each human LINE1 contains two regions from which gene products can be encoded. The first coding region contains a leucine zipper protein involved in protein-protein interactions and a protein that binds to the terminus of

nucleic acids. The second coding region has a purine/pyrimidine nuclease, reverse transcriptase and protein rich in amino acids cysteines and histidines. The end of the human LINE1, as with other retrotransposons is adenine-rich.

- Human L1 actively retrotransposes in the human genome. A recent study identified 1,708 somatic L1 retrotransposition events, especially in colorectal epithelial cells. These events occur from early embryogenesis and retrotransposition rate is substantially increased during colorectal tumourigenesis.

- **SINEs:**
  - SINEs are much shorter (300bp) than LINEs. They share similarity with genes transcribed by RNA polymerase II, the enzyme that transcribes genes into mRNA transcripts, and the initiation sequence of RNA polymerase III, the enzyme that transcribes genes into ribosomal RNA, tRNA and other small RNA molecules. SINEs such as mammalian MIR elements have tRNA gene at the start and adenine-rich at the end like in LINEs.
  - SINEs do not encode a functional reverse transcriptase protein and rely on other mobile transposons, especially LINEs. SINEs exploit LINE transposition components despite LINE-binding proteins prefer binding to LINE RNA. SINEs cannot transpose by themselves because they cannot encode SINE transcripts. They usually consist of parts derived from tRNA and LINEs. The tRNA portion contains an RNA polymerase III promoter which the same kind of enzyme as RNA polymerase II. This makes sure the LINE copies would be transcribed into RNA for further transposition. The LINE component remains so LINE-binding proteins can recognise the LINE part of the SINE.

- **Alu Elements:**
  - *Alus* are the most common SINE in primates. They are approximately 350 base pairs long, do not encode proteins and can be recognized by the restriction enzyme AluI (hence the name). Their distribution may be important in some genetic diseases and cancers. Copy and pasting Alu RNA requires the Alu's adenine-rich end and the rest of the sequence bound to a signal. The signal-bound Alu can then associate with ribosomes. LINE RNA associates on the same ribosomes as the Alu. Binding to the same ribosome allows Alus of SINEs to interact with LINE. This simultaneous translation of Alu element and LINE allows SINE copy and pasting.

- **SVA Elements:**
  - SVA elements are present at lower levels than SINES and LINEs in humans. The starts of SVA and Alu elements are similar, followed by repeats and an end similar to endogenous retrovirus. LINEs bind to sites flanking SVA elements to transpose them. SVA are one of the youngest transposons in great apes genome and among the most active and polymorphic in the human

population. SVA was created by a fusion between an Alu element, a VNTR (variable number tandem repeat), and an LTR fragment.

Role in Human Disease

Retrotransposons ensure they are not lost by chance by occurring only in cell genetics that can be passed on from one generation to the next from parent gametes. However, LINEs can transpose into the human embryo cells that eventually develop into the nervous system, raising the question whether this LINE retrotransposition affects brain function. LINE retrotransposition is also a feature of several cancers, but it is unclear whether retrotransposition itself causes cancer instead of just a symptom. Uncontrolled retrotransposition is bad for both the host organism and retrotransposons themselves so they have to be regulated. Retrotransposons are regulated by RNA interference. RNA interference is carried out by a bunch of short non-coding RNAs. The short non-coding RNA interacts with protein Argonaute to degrade retrotransposon transcripts and change their DNA histone structure to reduce their transcription.

Role in Evolution

LTR retrotransposons came about later than non-LTR retrotransposons, possibly from an ancestral non-LTR retrotransposon acquiring an integrase from a DNA transposon. Retroviruses gained additional properties to their virus envelopes by taking the relevant genes from other viruses using the power of LTR retrotransposon.

Due to their retrotransposition mechanism, retrotransposons amplify in number quickly, composing 40% of the human genome. The insertion rates for LINE1, Alu and SVA elements are 1/200 – 1/20, 1/20 and 1/900 respectively. The LINE1 insertion rates have varied a lot over the past 35 million years, so they indicate points in genome evolution.

Notably a large number of 100 kilobases in the maize genome show variety due to the presence or absence of retrotransposons. However since maize is unusual genetically as compared to other plants it cannot be used to predict retrotransposition in other plants.

7.2.2 Class II: DNA Transposons

DNA transposons are DNA sequences, sometimes referred to "jumping genes", that can move and integrate to different locations within the genome. They are class II transposable elements (TEs) that move through a DNA intermediate, as opposed to class I TEs, retrotransposons, that move through an RNA intermediate. DNA transposons can move in the DNA of an organism via a single-or double-stranded DNA intermediate. DNA transposons have been found in both prokaryotic and eukaryotic organisms. They can make up a significant portion of an organism's genome, particularly in eukaryotes. In prokaryotes, TE's can facilitate the horizontal transfer of antibiotic resistance or other genes associated with virulence. After replicating and propagating in a host, all transposon copies become inactivated and are lost unless the transposon passes to a genome by starting a new life cycle with horizontal transfer. DNA transposons do not randomly insert themselves into the genome, but rather show preference for specific sites.

With regard to movement, DNA transposons can be categorized as autonomous and nonautonomous. Autonomous ones can move on their own, while nonautonomous ones require the presence of another transposable element's gene, transposase, to move. There are three main classifications for movement for DNA transposons: "cut and paste," "rolling circle" (Helitrons), and "self-synthesizing" (Polintons). These distinct mechanisms of movement allow them to move around the genome of an organism. Since DNA transposons cannot synthesize DNA, they replicate using the host replication machinery. These three main classes are then further broken down into 23 different superfamilies characterized by their structure, sequence, and mechanism of action.

DNA transposons are a cause of gene expression alterations. As newly inserted DNA into active coding sequences, they can disrupt normal protein functions and cause mutations. Class II TEs make up about 3% of the human genome. Today, there are no active DNA transposons in the human genome. Therefore, the elements found in the human genome are called fossils.

Mechanisms of Action

Cut and Paste

Traditionally, DNA transposons move around in the genome by a cut and paste method. The system requires a transposase enzyme that catalyzes the movement of the DNA from its current location in the genome and inserts it in a new location. Transposition requires three DNA sites on the transposon: two at each end of the transposon called terminal inverted repeats and one at the target site. The transposase will bind to the terminal inverted repeats of the transposon and mediate synapsis of the transposon ends. The transposase enzyme then disconnects the element from the flanking DNA of the original donor site and mediates the joining reaction that links the transposon to the new insertion site. The addition of the new DNA into the target site causes short gaps on either side of the inserted segment. Host systems repair these gaps resulting in the target sequence duplication (TSD) that are characteristic of transposition. In many reactions, the transposon is completely excised from the donor site in what is called a "cut and paste" transposition and inserted into the target DNA to form a simple insertion. Occasionally, genetic material not originally in the transposable element gets copied and moved as well.

Helitrons

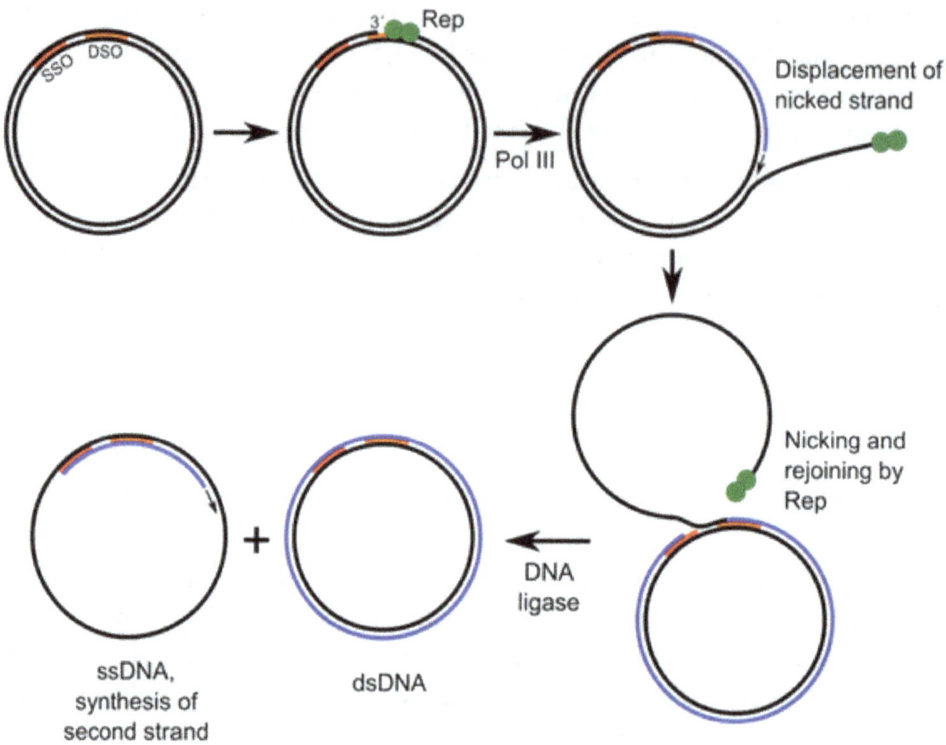

Fig: Rolling Circle Replication of a Circular DNA Plasmid

Helitrons are also a group of eukaryotic class II TEs. Helitrons do not follow the classical "cut and paste" mechanism. Instead, they are hypothesized to move around the genome via a rolling circle like mechanism. This process involves making a nick to a circular strand by an enzyme, which separates the DNA into two single strands.

The initiation protein then remains attached to the 5' Phosphate on the nicked strand, exposing the 3' hydroxyl of the complementary strand. This allows a polymerase enzyme to begin replication on the un-nicked strand.

Eventually the entire strand is replicated at which point the newly synthesized DNA disassociates and is replicated in parallel with the original template strand. Helitrons encode an unknown protein which is thought to have HUH endonuclease function as well as 5' to 3' helicase activity. This enzyme would make a single stranded cut in the DNA which explains the lack of Target Site Duplications found in Helitrons. Helitrons were also the first class of transposable elements to be discovered computationally and marked a paradigm shift in the way that whole genomes were studied.

Polintons

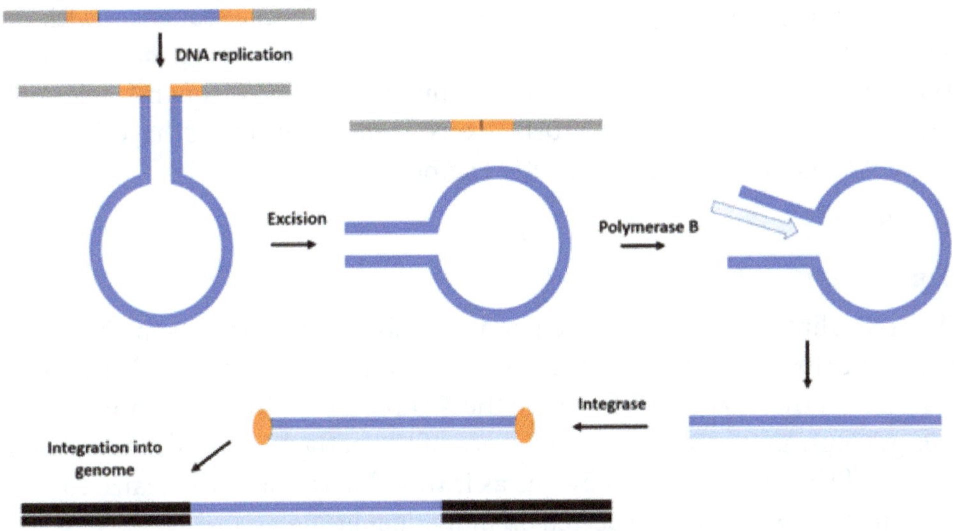

Fig: Self-Synthesizing Transposition Mechanism for Polintons

Polintons are also a group of eukaryotic class II TEs. As one of the most complex known DNA transposons in eukaryotes, they make up the genomes of protists, fungi, and animals, such as the entamoeba, soybean rust, and chicken, respectively. They contain genes with homology to viral proteins and which are often found in eukaryotic genomes, like polymerase and retroviral integrase. However, there is no known protein functionally similarly to the viral capsid or envelope proteins. They share their many structural characteristics with linear plasmids, bacteriophages and adenoviruses, which replicate using protein-primed DNA polymerases. Polintons have been proposed to go through a similar self-synthesis by their polymerase. Polintons, 15–20 kb long, encode up to 10 individual proteins. For replication, they utilize a protein-primed DNA polymerase B, retroviral integrase, cysteine protease, and ATPase. First, during host genome replication, a single-stranded extra-chromosomal Polinton element is excised from the host DNA using the integrase, forming a racket-like structure. Second, the Polinton undergoes replication using the DNA polymerase B, with initiation started by a terminal protein, which may encoded in some linear plasmids. Once the double stranded Polinton is generated, the integrase serves to insert it into the host genome. Polintons exhibit high variability between difference species and may tightly regulated, resulting in a low frequency rate in many genomes.

Effects of Transposons

DNA transposons, like all transposons, are quite impactful with respect to gene expression. A sequence of DNA may insert itself into a previously functional gene and create a mutation. This can happen in three distinct ways: 1. alteration of function, 2. chromosomal rearrangement, and 3. a source of novel genetic material. Since DNA transposons may happen to take parts of genomic sequences with them, exon shuffling may occur. Exon shuffling is the creation of novel gene products due to the new placement of two previously unrelated exons through transposition. Because of their ability to alter DNA expression, transposons have become an important target of research in genetic engineering.

Examples

Maize

Barbara McClintock first discovered and described DNA transposons in *Zea mays*, during the 1940s; this is an achievement that would earn her the Nobel Prize in 1983. She described the Ac/Ds system where the Ac unit (activator) was autonomous but the Ds genomic unit required the presence of the activator in order to move. This TE is one of the most visually obvious as it was able to cause the maize to change color from yellow to brown/spotted on individual kernels.

Fruit Flies

The Mariner/Tc1 transposon, found in many animals but studied in *Drosophila* was first described by Jacobson and Hartl. Mariner is well known for being able to excise and insert horizontally in to a new organism. Thousands of copies of the TE have been found interspersed in the human genome as well as other animals.

The Hobo transposons in Drosophila have been extensively studied due to their ability to cause gonadal dysgenesis. The insertion and subsequent expression of hobo-like sequences results in the loss of germ cells in the gonads of developing flies.

Bacteria

Bacterial transposons are especially good at facilitating horizontal gene transfer between microbes. Transposition facilitates the transfer and accumulation of antibiotic resistance genes. In bacteria, transposable elements can easily jump between the chromosomal genome and plasmids. In a 1982 study by Devaud et al., a multi-drug resistant strain of *Acinetobacter* was isolated and examined. Evidence pointed to the transfer of a plasmid in to the bacterium, where the resistance genes were transposed in to the chromosomal genome.

Genetic Diversity

Transposons may have an effect on the promotion of genetic diversity of many organisms. DNA transposons can drive the evolution of genomes by promoting the relocation of sections of DNA sequences. As a result, this can alter gene regulatory regions and phenotypes. The discovery of transposons was made by Barbara McClintock who noticed that these elements could actually change the color of the maize plants she was studying, providing quick evidence of one outcome from transposon movement. Another example is the Tol2 DNA transposon in medaka fish that is said to be the result of their variety in pigmentation patterns. These examples show that transposons can greatly influence the process of evolution by rapidly inducing changes in the genome.

Inactivation

All DNA transposons are inactive in the human genome. Inactivated, or silenced, transposons do not result in a phenotypic outcome and do not move around in the genome. Some are inactive because they have mutations that affect their ability to move between chromosomes, while others are capable of moving but remain inactive due to epigenetic defenses, like DNA methylation and chromatin remodeling. For example, chemical modifications of DNA can constrict certain areas of the genome such that transcription enzymes are unable to reach them. RNAi, specifically siRNA and miRNA silencing, is a naturally occurring mechanisms that, in addition to regulating eukaryotic gene expression, prevents transcription of DNA transposons. Another mode of inactivation is overproduction inhibition. When transposase exceeds a threshold concentration, transposon activity is decreased. Since transposase can form inactive or less active monomers that will decrease transposition activity overall, a decrease in the production of transposase will also occur when large copies of those less active elements increase in the host genome.

Horizontal Transfer

Horizontal transfer refers to the movement of DNA information between cells of different organisms. Horizontal transfer can involve the movement of TEs from one organism into the genome of another. The insertion itself allows the TE to become an activated gene in the new host. Horizontal transfer is used by DNA transposons to prevent inactivation and complete loss of the transposon. This inactivation is termed vertical inactivation, meaning that the DNA transposon is inactive and remains as a fossil. This type of transfer is not the most common, but has been seen in the case of the wheat virulence protein ToxA, which was transferred between the different fungal pathogens *Parastagonospora nodorum*, *Pyrenophora tritici-repentis*, and *Bipolaris sorokiniana*. Other examples include transfer between marine crustaceans, insects of different orders, and organisms of different phyla, such as humans and nematodes.

Evolution

Eukaryotic genomes differ in TE content. Recently, a study of the different superfamilies of TEs reveals that there are striking similarities between the groups. It has been hypothesized that many of them are represented in two or more Eukaryotic supergroups. This means that divergence of the transposon superfamilies could even predate the divergence of Eukaryotic supergroups.

V(D)J Recombination

V(D)J recombination, although not a DNA TE, is remarkably similar to transposons. V(D)J recombination is the process by which the large variation in antibody binding sites is created. In this mechanism, DNA is recombined in order to create genetic diversity.

Because of this, it has been hypothesized that these proteins, particularly *Rag1* and *Rag2* are derived from transposable elements.

Extinction in the Human Genome

There is evidence suggesting that at least 40 human DNA transposon families were active during mammalian radiation and early primate lineage. Then, there was a pause in transpositional activity during the later portion of primate radiation, with a complete halt in transposon movement in an anthropoid primate ancestor. There is no evidence of any transposable element younger than about 37 million years.

7.3 MECHANISMS OF TRANSPOSITION

The process by which mobile DNA elements move within the genome depends on whether they are retrotransposons or DNA transposons.

- **Retrotransposition (Class I elements)** involves transcription of the element into RNA, followed by reverse transcription into DNA by a reverse transcriptase enzyme. This newly synthesized DNA is integrated into the genome at a different location, allowing retrotransposons to increase their copy number without removing the original element.
- **Transposition (Class II elements)** entails the excision of the DNA transposon from its initial location and reinsertion at another site. The transposase enzyme facilitates this movement by recognizing and binding to the terminal repeats of the transposon, catalyzing its removal and integration into a new genomic site.

7.4 FUNCTIONS OF MOBILE DNA ELEMENTS

Although transposable elements were initially considered "junk" DNA, studies have shown that they play various important roles in genomes. They contribute to genetic diversity and have an impact on gene regulation, genome evolution, and the creation of new genetic material.

- **Genome Evolution and Innovation:** Mobile DNA sequences drive evolution by introducing variations in the genome. By moving between locations, they can duplicate, disrupt, or rearrange genetic material, creating opportunities for new genes or regulatory elements to evolve.

- **Gene Regulation:** Transposable elements can influence gene expression by carrying regulatory sequences that affect nearby genes. For instance, LINE and SINE elements in mammals sometimes contain promoter sequences, allowing the host gene to adopt new regulatory features.
- **Formation of New Genes:** Transposable elements can facilitate the formation of new genes through exon shuffling and gene duplication. This mechanism allows for the combination of different gene segments, leading to the evolution of novel proteins with unique functions.
- **Stress Response:** Some organisms activate mobile DNA elements in response to stress, such as environmental challenges or pathogen attacks. By increasing genetic diversity, this activation may allow for better adaptation and survival under changing conditions.

7.5 IMPACT OF MOBILE DNA ON GENOME STABILITY

While mobile DNA elements can be beneficial, their ability to insert randomly into the genome poses potential risks. Their movement can disrupt essential genes, cause chromosomal rearrangements, and lead to mutations. For example:

- **Gene Disruption:** When transposable elements insert into or near a gene, they can interfere with its normal function. In humans, some genetic disorders, such as hemophilia and certain forms of cancer, have been associated with TE insertions that disrupt genes critical for normal cell function.
- **Chromosomal Instability:** TEs can promote chromosomal rearrangements, including inversions, duplications, and deletions. Such rearrangements can destabilize genomes, leading to diseases and developmental abnormalities.
- **Mutagenesis:** The insertion of a mobile DNA element into a gene can cause mutations, which may have harmful consequences or contribute to the development of diseases. For example, LINE-1 insertions have been implicated in some forms of cancer.

7.6 APPLICATIONS OF MOBILE DNA IN BIOTECHNOLOGY AND MEDICINE

Scientists have harnessed mobile DNA elements for various biotechnological and therapeutic applications, particularly in gene therapy, genetic engineering, and functional genomics.

- **Gene Therapy:** DNA transposons, such as Sleeping Beauty and PiggyBac, are used as tools to deliver therapeutic genes into patient cells. These transposons can insert genes into target cells, offering a potential treatment for genetic disorders.
- **Genomic Research:** Mobile DNA elements are used in "insertional mutagenesis" studies, where they are inserted randomly in the genome to study gene functions. By analyzing the effects of insertion on cellular processes, researchers can better understand gene function and regulatory networks.
- **Biotechnology and Synthetic Biology:** Transposons are engineered for use in genetic circuits and synthetic biology applications. By controlling the movement of TEs, scientists can develop biological tools for building programmable genetic constructs and synthetic organisms.

7.7 THE ROLE OF MOBILE DNA IN HUMAN EVOLUTION

Mobile DNA elements have profoundly influenced human evolution, contributing to genome structure, genetic diversity, and adaptation. Retrotransposons, in particular, have helped shape the human genome. Many human genes and regulatory elements contain remnants of TEs, suggesting that mobile DNA played a significant role in shaping our biology.

The insertion and propagation of TEs have added complexity to the human genome, promoting the evolution of new gene regulatory networks. For instance, some TEs contain hormone-responsive elements, which can influence gene expression in response to environmental factors. This adaptability likely contributed to evolutionary advantages in early humans, enabling responses to changing environments.

7.8 CONCLUSION

Mobile DNA sequences are dynamic and integral components of genomes that have influenced genetic evolution, diversity, and adaptability across various organisms. Despite initial misconceptions that labeled them as "junk," these mobile elements are now recognized as essential to the architecture and function of genomes. With their ability to move, duplicate, and restructure genetic material, transposable elements drive evolution, impact gene regulation, and, in some cases, play a role in the development of genetic diseases. The study of mobile DNA sequences has opened new doors for genetic research, biotechnology, and medicine, offering insights into genome function and providing tools for treating genetic disorders. The continued exploration of these "jumping genes" will deepen our understanding of genomes and unlock further potential in genetic engineering and therapeutic applications.

CHAPTER 8

EPIGENETIC MODIFICATIONS OF THE GENOME

8.1 INTRODUCTION TO EPIGENETICS

In biology, epigenetics is the study of heritable traits, or a stable change of cell function, that happen without changes to the DNA sequence. Epigenetics usually involves a change that is not erased by cell division, and affects the regulation of gene expression. Such effects on cellular and physiological phenotypic traits may result from environmental factors, or be part of normal development. Epigenetic factors can also lead to cancer.

The term also refers to the mechanism of changes: functionally relevant alterations to the genome that do not involve mutation of the nucleotide sequence. Examples of mechanisms that produce such changes are DNA methylation and histone modification, each of which alters how genes are expressed without altering the underlying DNA sequence. Further, non-coding RNA sequences have been shown to play a key role in the regulation of gene expression. Gene expression can be controlled through the action of repressor proteins that attach to silencer regions of the DNA. These epigenetic changes may last through cell divisions for the duration of the cell's life, and may also last for multiple generations, even though they do not involve changes in the underlying DNA sequence of the organism; instead, non-genetic factors cause the organism's genes to behave (or "express themselves") differently.

One example of an epigenetic change in eukaryotic biology is the process of cellular differentiation. During morphogenesis, totipotent stem cells become the various pluripotent cell lines of the embryo, which in turn become fully differentiated cells. In other words, as a single fertilized egg cell – the zygote – continues to divide, the resulting daughter cells change into all the different cell types in an organism, including neurons, muscle cells, epithelium, endothelium of blood vessels, etc., by activating some genes while inhibiting the expression of others.

8.1.1 Definitions

The term *epigenesis* has a generic meaning of "extra growth" that has been used in English since the 17th century. In scientific publications, the term *epigenetics* started to appear in the 1930s. However, its contemporary meaning emerged only in the 1990s.

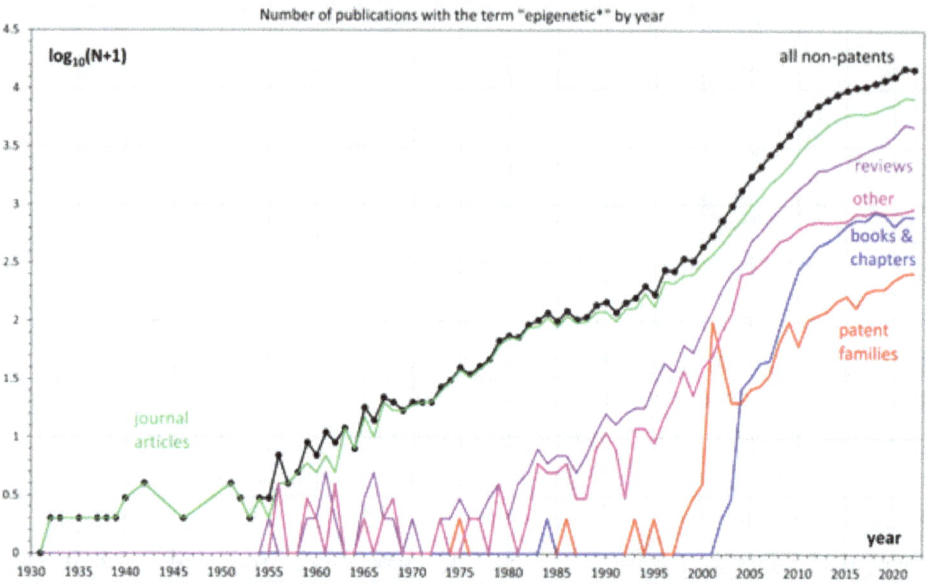

Fig: Number of Patent Families and Non-Patent Documents

A definition of the concept of *epigenetic trait* as a "stably heritable phenotype resulting from changes in a chromosome without alterations in the DNA sequence" was formulated at a Cold Spring Harbor meeting in 2008, although alternate definitions that include non-heritable traits are still being used widely.

Waddington's Canalisation, 1940s

The hypothesis of epigenetic changes affecting the expression of chromosomes was put forth by the Russian biologist Nikolai Koltsov. From the generic meaning, and the associated adjective *epigenetic*, British embryologist C. H. Waddington coined the term *epigenetics* in 1942 as pertaining to *epigenesis*, in parallel to Valentin Haecker's 'phenogenetics' (*Phänogenetik*). *Epigenesis* in the context of the biology of that period referred to the differentiation of cells from their initial totipotent state during embryonic development.

When Waddington coined the term, the physical nature of genes and their role in heredity was not known. He used it instead as a conceptual model of how genetic components might interact with their surroundings to produce a phenotype; he used the phrase "epigenetic landscape" as a metaphor for biological development. Waddington held that cell fates were established during development in a process he called canalisation much as a marble rolls down to the point of lowest local elevation. Waddington suggested visualising increasing irreversibility of cell type differentiation as ridges rising between the valleys where the marbles (analogous to cells) are travelling. In recent times, Waddington's notion of the epigenetic landscape has been rigorously formalized in the context of the systems dynamics state approach to the study of cell-fate. Cell-fate determination is predicted to exhibit certain dynamics, such as attractor-convergence (the attractor can be an equilibrium point, limit cycle or strange attractor) or oscillatory.

Contemporary

Robin Holliday defined in 1990 epigenetics as "the study of the mechanisms of temporal and spatial control of gene activity during the development of complex organisms."

More recent usage of the word in biology follows stricter definitions. As defined by Arthur Riggs and colleagues, it is "the study of mitotically and/or meiotically heritable changes in gene function that cannot be explained by changes in DNA sequence."

The term has also been used, however, to describe processes which have not been demonstrated to be heritable, such as some forms of histone modification. Consequently, there are attempts to redefine "epigenetics" in broader terms that would avoid the constraints of requiring heritability. For example, Adrian Bird defined epigenetics as "the structural adaptation of chromosomal regions so as to register, signal or perpetuate altered activity states." This definition would be inclusive of transient modifications associated with DNA repair or cell-cycle phases as well as stable changes maintained across multiple cell generations, but exclude others such as templating of membrane architecture and prions unless they impinge on chromosome function.

Such redefinitions however are not universally accepted and are still subject to debate. The NIH "Roadmap Epigenomics Project", which ran from 2008 to 2017, uses the following definition: "For purposes of this program, epigenetics refers to both heritable changes in gene activity and expression (in the progeny of cells or of individuals) and also stable, long-term alterations in the transcriptional potential of a cell that are not necessarily heritable." In 2008, a consensus definition of the epigenetic trait, a "stably heritable phenotype resulting from changes in a chromosome without alterations in the DNA sequence," was made at a Cold Spring Harbor meeting.

The similarity of the word to "genetics" has generated many parallel usages. The "epigenome" is a parallel to the word "genome", referring to the overall epigenetic state of a cell, and epigenomics refers to global analyses of epigenetic changes across the entire genome. The phrase "genetic code" has also been adapted – the "epigenetic code" has been used to describe the set of epigenetic features that create different phenotypes in different cells from the same underlying DNA sequence. Taken to its extreme, the "epigenetic code" could represent the total state of the cell, with the position of each molecule accounted for in an *epigenomic map*, a diagrammatic representation of the gene expression, DNA methylation and histone modification status of a particular genomic region. More typically, the term is used in reference to systematic efforts to measure specific, relevant forms of epigenetic information such as the histone code or DNA methylation patterns.

8.2 TYPES OF EPIGENETIC MODIFICATIONS

Epigenetic modifications mainly include DNA methylation, histone modifications, and RNA-associated mechanisms. Each of these modifications contributes uniquely to regulating gene expression by altering chromatin structure and accessibility, which determines whether genes are "turned on" or "off."

8.2.1 DNA Methylation

DNA methylation is one of the most studied forms of epigenetic modification. It involves the addition of a methyl group to the cytosine base in a DNA molecule, typically at CpG sites—regions of DNA where a cytosine nucleotide is followed by a guanine nucleotide. DNA methylation generally represses gene expression by altering the binding affinity of transcription factors or by recruiting proteins that compact chromatin, making it inaccessible to transcription machinery. This modification plays a critical role in processes such as X-chromosome inactivation, genomic imprinting, and silencing repetitive DNA sequences, which helps maintain genomic stability. Abnormal DNA methylation patterns are linked to numerous diseases, including cancer and neurological disorders.

8.2.2 Histone Modifications

Histones are proteins around which DNA wraps to form nucleosomes, the basic unit of chromatin. Histone modifications, which include methylation, acetylation, phosphorylation, and ubiquitination, occur on specific amino acid residues, mainly on the histone tails protruding from the nucleosome. These modifications influence chromatin structure, either loosening it to allow transcription (euchromatin) or tightening it to inhibit transcription (heterochromatin). For instance, acetylation of histones typically leads to gene activation by loosening chromatin structure, making it more accessible to transcriptional machinery. In contrast, histone methylation can either activate or repress genes depending on the specific amino acid residue modified. Together, histone modifications constitute a "histone code" that determines chromatin state and, consequently, gene expression.

8.2.3 Non-Coding RNA and RNA Interference

Non-coding RNAs (ncRNAs) also play a role in epigenetic regulation. MicroRNAs (miRNAs), small interfering RNAs (siRNAs), and long non-coding RNAs (lncRNAs) can modulate gene expression at the transcriptional and post-transcriptional levels. For instance, miRNAs bind to messenger RNA (mRNA) and prevent its translation, thereby inhibiting protein synthesis. In addition, some lncRNAs are involved in recruiting chromatin-modifying complexes to specific genomic regions, thereby influencing DNA methylation and histone modification patterns. The interplay between ncRNAs and other epigenetic mechanisms underscores the complexity of gene regulation within cells.

8.3 MECHANISMS OF EPIGENETIC CONTROL

The epigenome, which consists of the totality of epigenetic marks across an organism's genome, is established and maintained through a complex interplay of enzymes and other molecular machinery.

8.3.1 Writers, Erasers, and Readers

Epigenetic modifications are dynamically regulated by enzymes commonly referred to as "writers," "erasers," and "readers." Writers, such as DNA methyltransferases (DNMTs) and histone acetyltransferases (HATs), add epigenetic marks to the genome. Erasers, like histone deacetylases (HDACs) and demethylases, remove these marks, allowing for reversibility in epigenetic modifications. Readers are proteins that recognize and bind to specific epigenetic marks, interpreting them and recruiting other molecules to modify gene expression. For example, methyl-CpG-binding domain proteins bind to methylated DNA and attract repressive complexes that help maintain gene silencing. This network of enzymes and binding proteins ensures precise control over gene expression in response to environmental and developmental signals.

8.3.2 Chromatin Remodeling

Chromatin remodeling complexes alter the positioning of nucleosomes, which are histone-DNA complexes, to make certain regions of the genome more or less accessible to transcription factors. These complexes can slide nucleosomes along the DNA, evict histones entirely, or incorporate variant histones to modify chromatin structure. By changing nucleosome positioning, chromatin remodeling complexes work alongside epigenetic modifications to enable or inhibit transcription in specific genomic regions, adding another layer of gene expression control.

8.4 MOLECULAR BASIS

Epigenetic changes modify the activation of certain genes, but not the genetic code sequence of DNA. The microstructure (not code) of DNA itself or the associated chromatin proteins may be modified, causing activation or silencing. This mechanism enables differentiated cells in a multicellular organism to express only the genes that are necessary for their own activity. Epigenetic changes are preserved when cells divide. Most epigenetic changes only occur within the course of one individual organism's lifetime; however, these epigenetic changes can be transmitted to the organism's offspring through a process called transgenerational epigenetic inheritance. Moreover, if gene inactivation occurs in a sperm or egg cell that results in fertilization, this epigenetic modification may also be transferred to the next generation.

Specific epigenetic processes include paramutation, bookmarking, imprinting, gene silencing, X chromosome inactivation, position effect, DNA methylation reprogramming, transvection, maternal effects, the progress of carcinogenesis, many effects of teratogens, regulation of histone modifications and heterochromatin, and technical limitations affecting parthenogenesis and cloning.

8.4.1 DNA Damage

DNA damage can also cause epigenetic changes. DNA damage is very frequent, occurring on average about 60,000 times a day per cell of the human body. These damages are largely repaired, however, epigenetic changes can still remain at the site of DNA repair. In particular, a double strand break in DNA can initiate unprogrammed epigenetic gene silencing both by causing DNA methylation as well as by promoting silencing types of histone modifications (chromatin remodeling). In addition, the enzyme Parp1 (poly(ADP)-ribose polymerase) and its product poly(ADP)-ribose (PAR) accumulate at sites of DNA damage as part of the repair process. This accumulation, in turn, directs recruitment and activation of the chromatin remodeling protein, ALC1, that can cause nucleosome remodeling. Nucleosome remodeling has been found to cause, for instance, epigenetic silencing of DNA repair gene MLH1. DNA damaging chemicals, such as benzene, hydroquinone, styrene, carbon tetrachloride and trichloroethylene, cause considerable hypomethylation of DNA, some through the activation of oxidative stress pathways.

Foods are known to alter the epigenetics of rats on different diets. Some food components epigenetically increase the levels of DNA repair enzymes such as MGMT and MLH1 and p53. Other food components can reduce DNA damage, such as soy isoflavones. In one study, markers for oxidative stress, such as modified nucleotides that can result from DNA damage, were decreased by a 3-week diet supplemented with soy. A decrease in oxidative DNA damage was also observed 2 h after consumption of anthocyanin-rich bilberry (*Vaccinium myrtillius* L.) pomace extract.

8.4.2 DNA Repair

Damage to DNA is very common and is constantly being repaired. Epigenetic alterations can accompany DNA repair of oxidative damage or double-strand breaks. In human cells, oxidative DNA damage occurs about 10,000 times a day and DNA double-strand breaks occur about 10 to 50 times a cell cycle in somatic replicating cells. The selective advantage of DNA repair is to allow the cell to survive in the face of DNA damage. The selective advantage of epigenetic alterations that occur with DNA repair is not clear.

Repair of Oxidative DNA Damage can Alter Epigenetic Markers

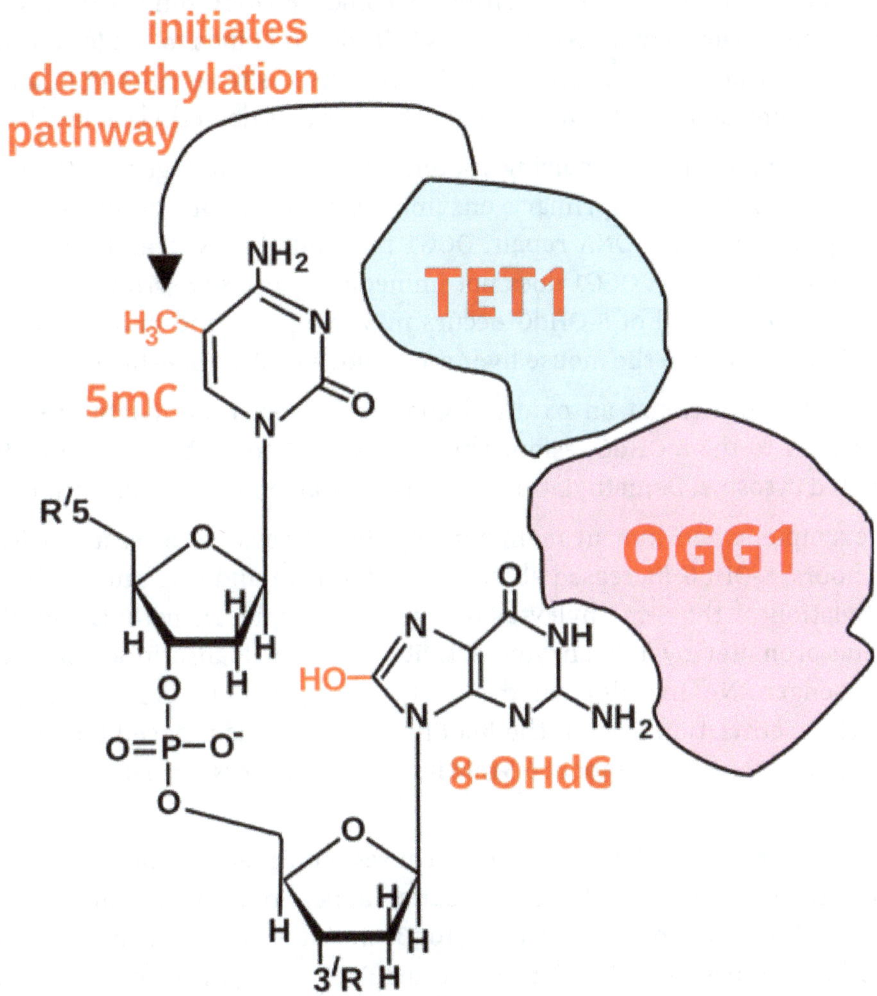

Fig: Initiation of DNA Demethylation at a CpG Site

In the steady state (with endogenous damages occurring and being repaired), there are about 2,400 oxidatively damaged guanines that form 8-oxo-2'-deoxyguanosine (8-OHdG) in the average mammalian cell DNA. 8-OHdG constitutes about 5% of the oxidative damages commonly present in DNA. The oxidized guanines do not occur randomly among all guanines in DNA. There is a sequence preference for the guanine at a methylated CpG site (a cytosine followed by guanine along its 5' → 3' direction and where the cytosine is methylated (5-mCpG)). A 5-mCpG site has the lowest ionization potential for guanine oxidation.

In adult somatic cells DNA methylation typically occurs in the context of CpG dinucleotides (CpG sites), forming 5-methylcytosine-pG, or 5mCpG. Reactive oxygen species (ROS) may attack guanine at the dinucleotide site, forming 8-hydroxy-2'-deoxyguanosine (8-OHdG), and resulting in a 5mCp-8-OHdG dinucleotide site. The base excision repair enzyme OGG1 targets 8-OHdG and binds to the lesion without immediate excision. OGG1, present at a 5mCp-8-OHdG site recruits TET1 and TET1 oxidizes the 5mC adjacent to the 8-OHdG. This initiates demethylation of 5mC.

Oxidized guanine has mispairing potential and is mutagenic. Oxoguanine glycosylase (OGG1) is the primary enzyme responsible for the excision of the oxidized guanine during DNA repair. OGG1 finds and binds to an 8-OHdG within a few seconds. However, OGG1 does not immediately excise 8-OHdG. In HeLa cells half maximum removal of 8-OHdG occurs in 30 minutes, and in irradiated mice, the 8-OHdGs induced in the mouse liver are removed with a half-life of 11 minutes.

When OGG1 is present at an oxidized guanine within a methylated CpG site it recruits TET1 to the 8-OHdG lesion. This allows TET1 to demethylate an adjacent methylated cytosine. Demethylation of cytosine is an epigenetic alteration.

As an example, when human mammary epithelial cells were treated with H_2O_2 for six hours, 8-OHdG increased about 3.5-fold in DNA and this caused about 80% demethylation of the 5-methylcytosines in the genome. Demethylation of CpGs in a gene promoter by TET enzyme activity increases transcription of the gene into messenger RNA. In cells treated with H_2O_2, one particular gene was examined, *BACE1*. The methylation level of the *BACE1* CpG island was reduced (an epigenetic alteration) and this allowed about 6.5 fold increase of expression of *BACE1* messenger RNA.

While six-hour incubation with H_2O_2 causes considerable demethylation of 5-mCpG sites, shorter times of H_2O_2 incubation appear to promote other epigenetic alterations. Treatment of cells with H_2O_2 for 30 minutes causes the mismatch repair protein heterodimer MSH2-MSH6 to recruit DNA methyltransferase 1 (DNMT1) to sites of some kinds of oxidative DNA damage. This could cause increased methylation of cytosines (epigenetic alterations) at these locations.

Jiang et al. treated HEK 293 cells with agents causing oxidative DNA damage, (potassium bromate (KBrO3) or potassium chromate (K2CrO4)). Base excision repair (BER) of oxidative damage occurred with the DNA repair enzyme polymerase beta localizing to oxidized guanines. Polymerase beta is the main human polymerase in short-patch BER of oxidative DNA damage. Jiang et al. also found that polymerase beta recruited the DNA methyltransferase protein DNMT3b to BER repair sites. They then evaluated the methylation pattern at the single nucleotide level in a small region of DNA including the promoter region and the early transcription region of the BRCA1 gene. Oxidative DNA damage from bromate modulated the DNA methylation pattern (caused epigenetic alterations) at CpG sites within the region of DNA studied. In untreated cells, CpGs located at −189, −134, −29, −19, +16, and +19 of the BRCA1 gene had methylated cytosines (where numbering is from the messenger RNA transcription start site, and negative numbers indicate nucleotides in the upstream promoter region). Bromate treatment-induced oxidation resulted in the loss of cytosine methylation at −189, −134, +16 and +19 while also leading to the formation of new methylation at the CpGs located at −80, −55, −21 and +8 after DNA repair was allowed.

Homologous Recombinational Repair Alters Epigenetic Markers

At least four articles report the recruitment of DNA methyltransferase 1 (DNMT1) to sites of DNA double-strand breaks. During homologous recombinational repair (HR) of the double-strand break, the involvement of DNMT1 causes the two repaired strands of DNA to have different levels of methylated cytosines. One strand becomes frequently methylated at about 21 CpG sites downstream of the repaired double-strand break. The other DNA strand loses methylation at about six CpG sites that were previously methylated downstream of the double-strand break, as well as losing methylation at about five CpG sites that were previously methylated upstream of the double-strand break. When the chromosome is replicated, this gives rise to one daughter chromosome that is heavily methylated downstream of the previous break site and one that is unmethylated in the region both upstream and downstream of the previous break site. With respect to the gene that was broken by the double-strand break, half of the progeny cells express that gene at a high level and in the other half of the progeny cells expression of that gene is repressed. When clones of these cells were maintained for three years, the new methylation patterns were maintained over that time period. In mice with a CRISPR-mediated homology-directed recombination insertion in their genome there were a large number of increased methylations of CpG sites within the double-strand break-associated insertion.

Non-Homologous end Joining can Cause Some Epigenetic Marker Alterations

Non-homologous end joining (NHEJ) repair of a double-strand break can cause a small number of demethylations of pre-existing cytosine DNA methylations downstream of the repaired double-strand break. Further work by Allen et al. showed that NHEJ of a DNA double-strand break in a cell could give rise to some progeny cells having repressed expression of the gene harboring the initial double-strand break and some progeny having high expression of that gene due to epigenetic alterations associated with NHEJ repair. The frequency of epigenetic alterations causing repression of a gene after an NHEJ repair of a DNA double-strand break in that gene may be about 0.9%.

8.4.3 Techniques Used to Study Epigenetics

Epigenetic research uses a wide range of molecular biological techniques to further understanding of epigenetic phenomena. These techniques include chromatin immunoprecipitation (together with its large-scale variants ChIP-on-chip and ChIP-Seq), fluorescent in situ hybridization, methylation-sensitive restriction enzymes, DNA adenine methyltransferase identification (DamID) and bisulfite sequencing. Furthermore, the use of bioinformatics methods has a role in computational epigenetics.

- **Chromatin Immunoprecipitation:** Chromatin Immunoprecipitation (ChIP) has helped bridge the gap between DNA and epigenetic interactions. With the use of ChIP, researchers are able to make findings in regards to gene regulation, transcription mechanisms, and chromatin structure.
- **Fluorescent *in situ* Hybridization:** Fluorescent *in situ* hybridization (FISH) is very important to understand epigenetic mechanisms. FISH can be used to find the location of genes on chromosomes, as well as finding noncoding RNAs. FISH is predominantly used for detecting chromosomal abnormalities in humans.
- **Methylation-Sensitive Restriction Enzymes:** Methylation sensitive restriction enzymes paired with PCR is a way to evaluate methylation in DNA - specifically the CpG sites. If DNA is methylated, the restriction enzymes will not cleave the strand. Contrarily, if the DNA is not methylated, the enzymes will cleave the strand and it will be amplified by PCR.
- **Bisulfite Sequencing:** Bisulfite sequencing is another way to evaluate DNA methylation. Cytosine will be changed to uracil from being treated with sodium bisulfite, whereas methylated cytosines will not be affected.
- **Nanopore Sequencing:** Certain sequencing methods, such as nanopore sequencing, allow sequencing of native DNA. Native (=unamplified) DNA retains the epigenetic modifications which would otherwise be lost during the amplification step. Nanopore basecaller models can distinguish between the signals obtained for epigenetically modified bases and unaltered based and provide an epigenetic profile in addition to the sequencing result.

8.4.4 Structural Inheritance

In ciliates such as *Tetrahymena* and *Paramecium*, genetically identical cells show heritable differences in the patterns of ciliary rows on their cell surface. Experimentally altered patterns can be transmitted to daughter cells. It seems existing structures act as templates for new structures. The mechanisms of such inheritance are unclear, but reasons exist to assume that multicellular organisms also use existing cell structures to assemble new ones.

8.4.5 Nucleosome Positioning

Eukaryotic genomes have numerous nucleosomes. Nucleosome position is not random, and determine the accessibility of DNA to regulatory proteins. Promoters active in different tissues have been shown to have different nucleosome positioning features. This determines differences in gene expression and cell differentiation. It has been shown that at least some nucleosomes are retained in sperm cells (where most but not all histones are replaced by protamines). Thus nucleosome positioning is to some degree inheritable. Recent studies have uncovered connections between nucleosome positioning and other epigenetic factors, such as DNA methylation and hydroxymethylation.

8.4.6 Histone Variants

Different histone variants are incorporated into specific regions of the genome non-randomly. Their differential biochemical characteristics can affect genome functions via their roles in gene regulation, and maintenance of chromosome structures.

8.4.7 Genomic Architecture

The three-dimensional configuration of the genome (the 3D genome) is complex, dynamic and crucial for regulating genomic function and nuclear processes such as DNA replication, transcription and DNA-damage repair.

8.5 FUNCTIONS AND CONSEQUENCES

8.5.1 In the Brain

Memory

Memory formation and maintenance are due to epigenetic alterations that cause the required dynamic changes in gene transcription that create and renew memory in neurons. An event can set off a chain of reactions that result in altered methylations of a large set of genes in neurons, which give a representation of the event, a memory. When a strong memory is created, as in a rat subjected to contextual fear conditioning (CFC), one of the earliest events to occur is that more than 100 DNA double-strand breaks are formed by topoisomerase IIB in neurons of the hippocampus and the medial prefrontal cortex (mPFC). These double-strand breaks are at specific locations that allow activation of transcription of immediate early genes (IEGs) that are important in memory formation, allowing their expression in mRNA, with peak mRNA transcription at seven to ten minutes after CFC.

Two important IEGs in memory formation are *EGR1* and the alternative promoter variant of *DNMT3A*, *DNMT3A2*. EGR1 protein binds to DNA at its binding motifs, 5'-GCGTGGGCG-3' or 5'-GCGGGGGCGG-3', and there are about 12,000 genome locations at which EGR1 protein can bind. EGR1 protein binds to DNA in gene promoter and enhancer regions. EGR1 recruits the demethylating enzyme TET1 to an association, and brings TET1 to about 600 locations on the genome where TET1 can then demethylate and activate the associated genes.

The DNA methyltransferases DNMT3A1, DNMT3A2 and DNMT3B can all methylate cytosines at CpG sites in or near the promoters of genes. As shown by Manzo et al., these three DNA methyltransferases differ in their genomic binding locations and DNA methylation activity at different regulatory sites. Manzo et al. located 3,970 genome regions exclusively enriched for DNMT3A1, 3,838 regions for DNMT3A2 and 3,432 regions for DNMT3B. When DNMT3A2 is newly induced as an IEG (when neurons are activated), many new cytosine methylations occur, presumably in the target regions of DNMT3A2. Oliviera et al. found that the neuronal activity-inducible IEG levels of Dnmt3a2 in the hippocampus determined the ability to form long-term memories.

Rats form long-term associative memories after contextual fear conditioning (CFC). Duke et al. found that 24 hours after CFC in rats, in hippocampus neurons, 2,097 genes (9.17% of the genes in the rat genome) had altered methylation. When newly methylated cytosines are present in CpG sites in the promoter regions of genes, the genes are often repressed, and when newly demethylated cytosines are present the genes may be activated. After CFC, there were 1,048 genes with reduced mRNA expression and 564 genes with upregulated mRNA expression. Similarly, when mice undergo CFC, one hour later in the hippocampus region of the mouse brain there are 675 demethylated genes and 613 hypermethylated genes. However, memories do not remain in the hippocampus, but after four or five weeks the memories are stored in the anterior cingulate cortex. In the studies on mice after CFC, Halder et al. showed that four weeks after CFC there were at least 1,000 differentially methylated genes and more than 1,000 differentially expressed genes in the anterior cingulate cortex, while at the same time the altered methylations in the hippocampus were reversed.

The epigenetic alteration of methylation after a new memory is established creates a different pool of nuclear mRNAs. As reviewed by Bernstein, the epigenetically determined new mix of nuclear mRNAs are often packaged into neuronal granules, or messenger RNP, consisting of mRNA, small and large ribosomal subunits, translation initiation factors and RNA-binding proteins that regulate mRNA function. These neuronal granules are transported from the neuron nucleus and are directed, according to 3' untranslated regions of the mRNA in the granules (their "zip codes"), to neuronal dendrites. Roughly 2,500 mRNAs may be localized to the dendrites of hippocampal pyramidal neurons and perhaps 450 transcripts are in excitatory presynaptic nerve terminals (dendritic spines). The altered assortments of transcripts (dependent on epigenetic alterations in the neuron nucleus) have different sensitivities in response to signals, which is the basis of altered synaptic plasticity. Altered synaptic plasticity is often considered the neurochemical foundation of learning and memory.

8.5.2 Development

Developmental epigenetics can be divided into predetermined and probabilistic epigenesis. Predetermined epigenesis is a unidirectional movement from structural development in DNA to the functional maturation of the protein. "Predetermined" here means that development is scripted and predictable. Probabilistic epigenesis on the other hand is a bidirectional structure-function development with experiences and external molding development.

Somatic epigenetic inheritance, particularly through DNA and histone covalent modifications and nucleosome repositioning, is very important in the development of multicellular eukaryotic organisms. The genome sequence is static (with some notable exceptions), but cells differentiate into many different types, which perform different functions, and respond differently to the environment and intercellular signaling. Thus, as individuals develop, morphogens activate or silence genes in an epigenetically heritable fashion, giving cells a memory. In mammals, most cells terminally differentiate, with only stem cells retaining the ability to differentiate into several cell types ("totipotency" and "multipotency").

In mammals, some stem cells continue producing newly differentiated cells throughout life, such as in neurogenesis, but mammals are not able to respond to loss of some tissues, for example, the inability to regenerate limbs, which some other animals are capable of. Epigenetic modifications regulate the transition from neural stem cells to glial progenitor cells (for example, differentiation into oligodendrocytes is regulated by the deacetylation and methylation of histones). Unlike animals, plant cells do not terminally differentiate, remaining totipotent with the ability to give rise to a new individual plant.

While plants do utilize many of the same epigenetic mechanisms as animals, such as chromatin remodeling, it has been hypothesized that some kinds of plant cells do not use or require "cellular memories", resetting their gene expression patterns using positional information from the environment and surrounding cells to determine their fate.

Epigenetic changes can occur in response to environmental exposure - for example, maternal dietary supplementation with genistein (250 mg/kg) have epigenetic changes affecting expression of the agouti gene, which affects their fur color, weight, and propensity to develop cancer. Ongoing research is focused on exploring the impact of other known teratogens, such as diabetic embryopathy, on methylation signatures.

Controversial results from one study suggested that traumatic experiences might produce an epigenetic signal that is capable of being passed to future generations. Mice were trained, using foot shocks, to fear a cherry blossom odor. The investigators reported that the mouse offspring had an increased aversion to this specific odor. They suggested epigenetic changes that increase gene expression, rather than in DNA itself, in a gene, M71, that governs the functioning of an odor receptor in the nose that responds specifically to this cherry blossom smell. There were physical changes that correlated with olfactory (smell) function in the brains of the trained mice and their descendants. Several criticisms were reported, including the study's low statistical power as evidence of some irregularity such as bias in reporting results.

Due to limits of sample size, there is a probability that an effect will not be demonstrated to within statistical significance even if it exists. The criticism suggested that the probability that all the experiments reported would show positive results if an identical protocol was followed, assuming the claimed effects exist, is merely 0.4%. The authors also did not indicate which mice were siblings, and treated all of the mice as statistically independent. The original researchers pointed out negative results in the paper's appendix that the criticism omitted in its calculations, and undertook to track which mice were siblings in the future.

8.5.3 Transgenerational Epigenetic Inheritance

Epigenetic mechanisms were a necessary part of the evolutionary origin of cell differentiation. Although epigenetics in multicellular organisms is generally thought to be a mechanism involved in differentiation, with epigenetic patterns "reset" when organisms reproduce, there have been some observations of transgenerational epigenetic inheritance (e.g., the phenomenon of paramutation observed in maize).

Although most of these multigenerational epigenetic traits are gradually lost over several generations, the possibility remains that multigenerational epigenetics could be another aspect to evolution and adaptation.

A sequestered germ line or Weismann barrier is specific to animals, and epigenetic inheritance is more common in plants and microbes. Eva Jablonka, Marion J. Lamb and Étienne Danchin have argued that these effects may require enhancements to the standard conceptual framework of the modern synthesis and have called for an extended evolutionary synthesis. Other evolutionary biologists, such as John Maynard Smith, have incorporated epigenetic inheritance into population-genetics models or are openly skeptical of the extended evolutionary synthesis (Michael Lynch). Thomas Dickins and Qazi Rahman state that epigenetic mechanisms such as DNA methylation and histone modification are genetically inherited under the control of natural selection and therefore fit under the earlier "modern synthesis".

Two important ways in which epigenetic inheritance can differ from traditional genetic inheritance, with important consequences for evolution, are:

- Rates of epimutation can be much faster than rates of mutation.
- The epimutations are more easily reversible.

In plants, heritable DNA methylation mutations are 100,000 times more likely to occur compared to DNA mutations. An epigenetically inherited element such as the PSI+ system can act as a "stop-gap", good enough for short-term adaptation that allows the lineage to survive for long enough for mutation and/or recombination to genetically assimilate the adaptive phenotypic change. The existence of this possibility increases the evolvability of a species.

More than 100 cases of transgenerational epigenetic inheritance phenomena have been reported in a wide range of organisms, including prokaryotes, plants, and animals. For instance, mourning-cloak butterflies will change color through hormone changes in response to experimentation of varying temperatures.

The filamentous fungus *Neurospora crassa* is a prominent model system for understanding the control and function of cytosine methylation. In this organism, DNA methylation is associated with relics of a genome-defense system called RIP (repeat-induced point mutation) and silences gene expression by inhibiting transcription elongation.

The yeast prion PSI is generated by a conformational change of a translation termination factor, which is then inherited by daughter cells. This can provide a survival advantage under adverse conditions, exemplifying epigenetic regulation which enables unicellular organisms to respond rapidly to environmental stress. Prions can be viewed as epigenetic agents capable of inducing a phenotypic change without modification of the genome.

Direct detection of epigenetic marks in microorganisms is possible with single molecule real time sequencing, in which polymerase sensitivity allows for measuring methylation and other modifications as a DNA molecule is being sequenced. Several projects have demonstrated the ability to collect genome-wide epigenetic data in bacteria.

8.6 EPIGENETICS IN BACTERIA

While epigenetics is of fundamental importance in eukaryotes, especially metazoans, it plays a different role in bacteria. Most importantly, eukaryotes use epigenetic mechanisms primarily to regulate gene expression which bacteria rarely do. However, bacteria make widespread use of postreplicative DNA methylation for the epigenetic control of DNA-protein interactions. Bacteria also use DNA adenine methylation (rather than DNA cytosine methylation) as an epigenetic signal. DNA adenine methylation is important in bacteria virulence in organisms such as *Escherichia coli, Salmonella, Vibrio, Yersinia, Haemophilus*, and *Brucella*. In *Alphaproteobacteria*, methylation of adenine regulates the cell cycle and couples gene transcription to DNA replication. In *Gammaproteobacteria*, adenine methylation provides signals for DNA replication, chromosome segregation, mismatch repair, packaging of bacteriophage, transposase activity and regulation of gene expression. There exists a genetic switch controlling *Streptococcus pneumoniae* (the pneumococcus) that allows the bacterium to randomly change its characteristics into six alternative states that could pave the way to improved vaccines. Each form is randomly generated by a phase variable methylation system. The ability of the pneumococcus to cause deadly infections is different in each of these six states. Similar systems exist in other bacterial genera. In Bacillota such as *Clostridioides difficile*, adenine methylation regulates sporulation, biofilm formation and host-adaptation.

8.7 EPIGENETIC MODIFICATIONS IN DEVELOPMENT AND DIFFERENTIATION

Epigenetic modifications are essential for normal development and cellular differentiation. During embryonic development, cells undergo successive rounds of division and specialization to form the various tissues and organs of an organism. Epigenetic mechanisms ensure that genes are activated or silenced as needed in different cell types. For instance, pluripotent stem cells have an open chromatin structure that allows them to express genes needed for all cell types. As differentiation proceeds, epigenetic marks selectively silence genes unnecessary for a particular cell type, leading to a stable and heritable cell identity. This specialization is maintained by the epigenetic landscape, which "locks in" gene expression patterns appropriate for each cell type, ensuring tissue function and stability.

8.8 ENVIRONMENTAL INFLUENCES ON THE EPIGENOME

The epigenome is highly responsive to environmental factors, including diet, stress, exposure to toxins, and physical activity. These external factors can induce changes in DNA methylation patterns, histone modifications, and ncRNA expression. For example, studies have shown that maternal diet during pregnancy can influence DNA methylation patterns in offspring, affecting their health outcomes later in life. Similarly, exposure to toxins like cigarette smoke or heavy metals has been linked to aberrant epigenetic modifications associated with diseases such as cancer and cardiovascular conditions. This adaptability highlights the role of the epigenome in mediating gene-environment interactions, allowing organisms to respond to changes in their surroundings dynamically.

8.9 EPIGENETICS AND DISEASE

Alterations in epigenetic patterns are implicated in various diseases, particularly cancer. In cancer cells, DNA methylation patterns often become abnormal, leading to the silencing of tumor suppressor genes and the activation of oncogenes. Similarly, histone modifications can be disrupted in cancer, resulting in an altered chromatin state that promotes uncontrolled cell proliferation. Epigenetic changes are also implicated in neurodegenerative diseases, cardiovascular conditions, and autoimmune disorders. The reversibility of epigenetic modifications makes them attractive therapeutic targets, as drugs can be designed to inhibit or modulate enzymes involved in adding or removing epigenetic marks.

8.10 MEDICINE

8.10.1 Twins

Direct comparisons of identical twins constitute an optimal model for interrogating environmental epigenetics. In the case of humans with different environmental exposures, monozygotic (identical) twins were epigenetically indistinguishable during their early years, while older twins had remarkable differences in the overall content and genomic distribution of 5-methylcytosine DNA and histone acetylation. The twin pairs who had spent less of their lifetime together and/or had greater differences in their medical histories were those who showed the largest differences in their levels of 5-methylcytosine DNA and acetylation of histones H3 and H4. Dizygotic (fraternal) and monozygotic (identical) twins show evidence of epigenetic influence in humans. DNA sequence differences that would be abundant in a singleton-based study do not interfere with the analysis. Environmental differences can produce long-term epigenetic effects, and different developmental monozygotic twin subtypes may be different with respect to their susceptibility to be discordant from an epigenetic point of view.

A high-throughput study, which denotes technology that looks at extensive genetic markers, focused on epigenetic differences between monozygotic twins to compare global and locus-specific changes in DNA methylation and histone modifications in a sample of 40 monozygotic twin pairs. In this case, only healthy twin pairs were studied, but a wide range of ages was represented, between 3 and 74 years. One of the major conclusions from this study was that there is an age-dependent accumulation of epigenetic differences between the two siblings of twin pairs.

This accumulation suggests the existence of epigenetic "drift". *Epigenetic drift* is the term given to epigenetic modifications as they occur as a direct function with age. While age is a known risk factor for many diseases, age-related methylation has been found to occur differentially at specific sites along the genome. Over time, this can result in measurable differences between biological and chronological age. Epigenetic changes have been found to be reflective of lifestyle and may act as functional biomarkers of disease before clinical threshold is reached.

A more recent study, where 114 monozygotic twins and 80 dizygotic twins were analyzed for the DNA methylation status of around 6000 unique genomic regions, concluded that epigenetic similarity at the time of blastocyst splitting may also contribute to phenotypic similarities in monozygotic co-twins. This supports the notion that microenvironment at early stages of embryonic development can be quite important for the establishment of epigenetic marks. Congenital genetic disease is well understood and it is clear that epigenetics can play a role, for example, in the case of Angelman syndrome and Prader–Willi syndrome. These are normal genetic diseases caused by gene deletions or inactivation of the genes but are unusually common because individuals are essentially hemizygous because of genomic imprinting, and therefore a single gene knock out is sufficient to cause the disease, where most cases would require both copies to be knocked out.

8.10.2 Genomic Imprinting

Some human disorders are associated with genomic imprinting, a phenomenon in mammals where the father and mother contribute different epigenetic patterns for specific genomic loci in their germ cells. The best-known case of imprinting in human disorders is that of Angelman syndrome and Prader–Willi syndrome - both can be produced by the same genetic mutation, chromosome 15q partial deletion, and the particular syndrome that will develop depends on whether the mutation is inherited from the child's mother or from their father.

In the Överkalix study, paternal (but not maternal) grandsons of Swedish men who were exposed during preadolescence to famine in the 19th century were less likely to die of cardiovascular disease. If food was plentiful, then diabetes mortality in the grandchildren increased, suggesting that this was a transgenerational epigenetic inheritance. The opposite effect was observed for females - the paternal (but not maternal) granddaughters of women who experienced famine while in the womb (and therefore while their eggs were being formed) lived shorter lives on average.

8.10.3 Examples of Drugs Altering Gene Expression from Epigenetic Events

The use of beta-lactam antibiotics can alter glutamate receptor activity and the action of cyclosporine on multiple transcription factors. Additionally, lithium can impact autophagy of aberrant proteins, and opioid drugs via chronic use can increase the expression of genes associated with addictive phenotypes.

Parental nutrition, in utero exposure to stress or endocrine disrupting chemicals, male-induced maternal effects such as the attraction of differential mate quality, and maternal as well as paternal age, and offspring gender could all possibly influence whether a germline epimutation is ultimately expressed in offspring and the degree to which intergenerational inheritance remains stable throughout posterity. However, whether and to what extent epigenetic effects can be transmitted across generations remains unclear, particularly in humans.

8.10.4 Addiction

Addiction is a disorder of the brain's reward system which arises through transcriptional and neuroepigenetic mechanisms and occurs over time from chronically high levels of exposure to an addictive stimulus (e.g., morphine, cocaine, sexual intercourse, gambling). Transgenerational epigenetic inheritance of addictive phenotypes has been noted to occur in preclinical studies. However, robust evidence in support of the persistence of epigenetic effects across multiple generations has yet to be established in humans; for example, an epigenetic effect of prenatal exposure to smoking that is observed in great-grandchildren who had not been exposed.

8.11 RESEARCH

The two forms of heritable information, namely genetic and epigenetic, are collectively called dual inheritance. Members of the APOBEC/AID family of cytosine deaminases may concurrently influence genetic and epigenetic inheritance using similar molecular mechanisms, and may be a point of crosstalk between these conceptually compartmentalized processes.

Fluoroquinolone antibiotics induce epigenetic changes in mammalian cells through iron chelation. This leads to epigenetic effects through inhibition of α-ketoglutarate-dependent dioxygenases that require iron as a co-factor.

Various pharmacological agents are applied for the production of induced pluripotent stem cells (iPSC) or maintain the embryonic stem cell (ESC) phenotypic via epigenetic approach. Adult stem cells like bone marrow stem cells have also shown a potential to differentiate into cardiac competent cells when treated with G9a histone methyltransferase inhibitor BIX01294.

Cell plasticity, which is the adaptation of cells to stimuli without changes in their genetic code, requires epigenetic changes. These have been observed in cell plasticity in cancer cells during epithelial-to-mesenchymal transition and also in immune cells, such as macrophages. Interestingly, metabolic changes underly these adaptations, since various metabolites play crucial roles in the chemistry of epigenetic marks.

This includes for instance alpha-ketoglutarate, which is required for histone demethylation, and acetyl-Coenzyme A, which is required for histone acetylation.

8.11.1 Epigenome Editing

Epigenetic regulation of gene expression that could be altered or used in epigenome editing are or include mRNA/lncRNA modification, DNA methylation modification and histone modification.

8.11.2 CpG Sites, SNPs and Biological Traits

Methylation is a widely characterized mechanism of genetic regulation that can determine biological traits. However, strong experimental evidences correlate methylation patterns in SNPs as an important additional feature for the classical activation/inhibition epigenetic dogma. Molecular interaction data, supported by colocalization analyses, identify multiple nuclear regulatory pathways, linking sequence variation to disturbances in DNA methylation and molecular and phenotypic variation.

UBASH3B Locus

UBASH3B encodes a protein with tyrosine phosphatase activity, which has been previously linked to advanced neoplasia. SNP rs7115089 was identified as influencing DNA methylation and expression of this locus, as well as and Body Mass Index (BMI). In fact, SNP rs7115089 is strongly associated with BMI and with genetic variants linked to other cardiovascular and metabolic traits in GWASs. New studies suggesting *UBASH3B* as a potential mediator of adiposity and cardiometabolic disease. In addition, animal models demonstrated that *UBASH3B* expression is an indicator of caloric restriction that may drive programmed susceptibility to obesity and it is associated with other measures of adiposity in human peripherical blood.

NFKBIE Locus

SNP rs730775 is located in the first intron of *NFKBIE* and is a *cis* eQTL for *NFKBIE* in whole blood. Nuclear factor (NF)-κB inhibitor ε (NFKBIE) directly inhibits NF-κB1 activity and is significantly co-expressed with NF-κB1, also, it is associated with rheumatoid arthritis. Colocalization analysis supports that variants for the majority of the CpG sites in SNP rs730775 cause genetic variation at the *NFKBIE* locus which is suggestible linked to rheumatoid arthritis through *trans* acting regulation of DNA methylation by NF-κB.

FADS1 Locus

Fatty acid desaturase 1 (FADS1) is a key enzyme in the metabolism of fatty acids. Moreover, rs174548 in the *FADS1* gene shows increased correlation with DNA methylation in people with high abundance of CD8+ T cells. SNP rs174548 is strongly associated with concentrations of arachidonic acid and other metabolites in fatty acid metabolism, blood eosinophil counts. and inflammatory diseases such as asthma. Interaction results indicated a correlation between rs174548 and asthma, providing new insights about fatty acid metabolism in CD8+ T cells with immune phenotypes.

8.12 PSEUDOSCIENCE

As epigenetics is in the early stages of development as a science and is surrounded by sensationalism in the public media, David Gorski and geneticist Adam Rutherford have advised caution against the proliferation of false and pseudoscientific conclusions by new age authors making unfounded suggestions that a person's genes and health can be manipulated by mind control. Misuse of the scientific term by quack authors has produced misinformation among the general public.

8.13 EPIGENETIC THERAPY AND FUTURE DIRECTIONS

Epigenetic therapy is an emerging field focused on modifying epigenetic marks to restore normal gene expression patterns in diseases. For example, DNMT inhibitors are used in treating certain cancers by reactivating silenced tumor suppressor genes. Similarly, HDAC inhibitors can change the acetylation patterns of histones, restoring normal gene expression in cancer cells. Research is also exploring ways to target specific histone modifications and ncRNA pathways to develop precise therapies for various diseases.

The potential for personalized epigenetic therapies is another promising direction. By profiling the unique epigenome of a patient, therapies can be tailored to target specific epigenetic abnormalities contributing to disease. However, the field faces challenges, including understanding the complex interactions within the epigenome and ensuring targeted delivery of epigenetic drugs to specific cells.

8.14 CONCLUSION

Epigenetic modifications are fundamental to gene regulation, organismal development, and response to environmental factors. These heritable changes, which do not alter the DNA sequence, play a central role in establishing and maintaining cellular identity and function. The reversibility of epigenetic marks opens exciting opportunities for therapeutic intervention in diseases where abnormal gene expression is a factor. As research continues, the understanding of epigenetics will likely expand, leading to innovative approaches to combat diseases, personalized medicine, and deeper insights into how organisms interact with and adapt to their environments.

Epigenetics exemplifies the intricate interplay between genetic and environmental influences, highlighting the complexity and adaptability of biological systems.

CHAPTER 9

RNA INTERFERENCE REGULATES THE GENOME

9.1 INTRODUCTION TO RNA INTERFERENCE

RNA interference (RNAi) is a natural, highly conserved biological process that regulates gene expression at the post-transcriptional level. Discovered in the 1990s, RNAi was initially observed in plants but is now known to be widespread across various organisms, including animals and humans. RNAi is a mechanism by which small RNA molecules inhibit gene expression or translation by neutralizing specific mRNA molecules. This powerful process plays essential roles in cellular defense, gene regulation, and genomic stability, making it a key area of research for understanding cellular mechanisms and developing genetic therapies.

9.2 MECHANISMS OF RNA INTERFERENCE

RNA interference operates through two primary small RNA molecules: microRNAs (miRNAs) and small interfering RNAs (siRNAs). These small RNAs differ in their origins but share a similar mechanism of action, utilizing protein complexes to bind and degrade target mRNA, effectively silencing specific genes.

9.2.1 MicroRNAs (miRNAs)

Micro ribonucleic acid (microRNA, miRNA, **μRNA**) are small, single-stranded, non-coding RNA molecules containing 21–23 nucleotides. Found in plants, animals, and even some viruses, miRNAs are involved in RNA silencing and post-transcriptional regulation of gene expression. miRNAs base-pair to complementary sequences in messenger RNA (mRNA) molecules, then silence said mRNA molecules by one or more of the following processes:

- Cleaving the mRNA strand into two pieces.
- Destabilizing the mRNA by shortening its poly(A) tail.
- Reducing translation of the mRNA into proteins.

In cells of humans and other animals, miRNAs primarily act by destabilizing the mRNA.

miRNAs resemble the small interfering RNAs (siRNAs) of the RNA interference (RNAi) pathway, except miRNAs derive from regions of RNA transcripts that fold back on themselves to form short hairpins, whereas siRNAs derive from longer regions of double-stranded RNA. The human genome may encode over 1900 miRNAs, However, only about 500 human miRNAs represent *bona fide* miRNAs in the manually curated miRNA gene database MirGeneDB.

miRNAs are abundant in many mammalian cell types. They appear to target about 60% of the genes of humans and other mammals. Many miRNAs are evolutionarily conserved, which implies that they have important biological functions. For example, 90 families of miRNAs have been conserved since at least the common ancestor of mammals and fish, and most of these conserved miRNAs have important functions, as shown by studies in which genes for one or more members of a family have been knocked out in mice.

In 2024, American scientists Victor Ambros and Gary Ruvkun were awarded the Nobel Prize in Physiology or Medicine for their work on the discovery of miRNA and its role in post-transcriptional gene regulation.

History

The first miRNA was discovered in the early 1990s. However, they were not recognized as a distinct class of biological regulators until the early 2000s. Research revealed different sets of miRNAs expressed in different cell types and tissues and multiple roles for miRNAs in plant and animal development and in many other biological processes. Aberrant miRNA expression are implicated in disease states. MiRNA-based therapies are under investigation.

The first miRNA was discovered in 1993 by a group led by Victor Ambros and including Lee and Feinbaum. However, additional insight into its mode of action required simultaneously published work by Gary Ruvkun's team, including Wightman and Ha. These groups published back-to-back papers on the *lin-4* gene, which was known to control the timing of *C. elegans* larval development by repressing the *lin-14* gene.

When Lee et al. isolated the *lin-4* miRNA, they found that instead of producing an mRNA encoding a protein, it produced short non-coding RNAs, one of which was a ~22-nucleotide RNA that contained sequences partially complementary to multiple sequences in the 3' UTR of the *lin-14* mRNA. This complementarity was proposed to inhibit the translation of the *lin-14* mRNA into the LIN-14 protein. At the time, the *lin-4* small RNA was thought to be a nematode idiosyncrasy.

In 2000, a second small RNA was characterized: *let-7* RNA, which represses *lin-41* to promote a later developmental transition in *C. elegans*. The *let-7* RNA was found to be conserved in many species, leading to the suggestion that *let-7* RNA and additional "small temporal RNAs" might regulate the timing of development in diverse animals, including humans.

A year later, the *lin-4* and *let-7* RNAs were found to be part of a large class of small RNAs present in *C. elegans, Drosophila* and human cells. The many RNAs of this class resembled the *lin-4* and *let-7* RNAs, except their expression patterns were usually inconsistent with a role in regulating the timing of development. This suggested that most might function in other types of regulatory pathways. At this point, researchers started using the term "microRNA" to refer to this class of small regulatory RNAs. The first human disease associated with deregulation of miRNAs was chronic lymphocytic leukemia. In this disorder, the miRNAs have a dual role working as both tumor suppressors and oncogenes.

Nomenclature

Under a standard nomenclature system, names are assigned to experimentally confirmed miRNAs before publication. The prefix "miR" is followed by a dash and a number, the latter often indicating order of naming. For example, miR-124 was named and likely discovered prior to miR-456. A capitalized "miR-" refers to the mature form of the miRNA, while the uncapitalized "mir-" refers to the pre-miRNA and the pri-miRNA. The genes encoding miRNAs are also named using the same three-letter prefix according to the conventions of the organism gene nomenclature. For examples, the official miRNAs gene names in some organisms are "*mir-1* in *C. elegans* and *Drosophila, Mir1* in *Rattus norvegicus* and *MIR25* in human.

Pre-miRNAs, pri-miRNAs and genes that lead to 100% identical mature miRNAs but that are located at different places in the genome are indicated with an additional dash-number suffix. For example, the pre-miRNAs hsa-mir-194-1 and hsa-mir-194-2 lead to an identical mature miRNA (hsa-miR-194) but are from genes located in different genome regions.

Species of origin is designated with a three-letter prefix, e.g., hsa-miR-124 is a human (*Homo sapiens*) miRNA and oar-miR-124 is a sheep (*Ovis aries*) miRNA. Other common prefixes include "v" for viral (miRNA encoded by a viral genome) and "d" for *Drosophila* miRNA (a fruit fly commonly studied in genetic research).

When two mature microRNAs originate from opposite arms of the same pre-miRNA and are found in roughly similar amounts, they are denoted with a -3p or -5p suffix. (In the past, this distinction was also made with "s" (sense) and "as" (antisense)). However, the mature microRNA found from one arm of the hairpin is usually much more abundant than that found from the other arm, in which case, an asterisk following the name indicates the mature species found at low levels from the opposite arm of a hairpin. For example, miR-124 and miR-124* share a pre-miRNA hairpin, but much more miR-124 is found in the cell.

Targets

Plant miRNAs usually have near-perfect pairing with their mRNA targets, which induces gene repression through cleavage of the target transcripts. In contrast, animal miRNAs are able to recognize their target mRNAs by using as few as 6–8 nucleotides (the seed region) at the 5' end of the miRNA, which is not enough pairing to induce cleavage of the target mRNAs. Combinatorial regulation is a feature of miRNA regulation in animals. A given miRNA may have hundreds of different mRNA targets, and a given target might be regulated by multiple miRNAs.

Estimates of the average number of unique messenger RNAs that are targets for repression by a typical miRNA vary, depending on the estimation method, but multiple approaches show that mammalian miRNAs can have many unique targets. For example, an analysis of the miRNAs highly conserved in vertebrates shows that each has, on average, roughly 400 conserved targets. Likewise, experiments show that a single miRNA species can reduce the stability of hundreds of unique messenger RNAs. Other experiments show that a single miRNA species may repress the production of hundreds of proteins, but that this repression often is relatively mild (much less than 2-fold).

Biogenesis

As many as 40% of miRNA genes may lie in the introns or even exons of other genes. These are usually, though not exclusively, found in a sense orientation, and thus usually are regulated together with their host genes. The DNA template is not the final word on mature miRNA production: 6% of human miRNAs show RNA editing (IsomiRs), the site-specific modification of RNA sequences to yield products different from those encoded by their DNA. This increases the diversity and scope of miRNA action beyond that implicated from the genome alone.

Transcription

miRNA genes are usually transcribed by RNA polymerase II (Pol II). The polymerase often binds to a promoter found near the DNA sequence, encoding what will become the hairpin loop of the pre-miRNA. The resulting transcript is capped with a specially modified nucleotide at the 5' end, polyadenylated with multiple adenosines (a poly(A) tail), and spliced. Animal miRNAs are initially transcribed as part of one arm of an ~80 nucleotide RNA stem-loop that in turn forms part of a several hundred nucleotide-long miRNA precursor termed a pri-miRNA. When a stem-loop precursor is found in the 3' UTR, a transcript may serve as a pri-miRNA and a mRNA. RNA polymerase III (Pol III) transcribes some miRNAs, especially those with upstream Alu sequences, transfer RNAs (tRNAs), and mammalian wide interspersed repeat (MWIR) promoter units.

Nuclear Processing

A single pri-miRNA may contain from one to six miRNA precursors. These hairpin loop structures are composed of about 70 nucleotides each. Each hairpin is flanked by sequences necessary for efficient processing.

The double-stranded RNA (dsRNA) structure of the hairpins in a pri-miRNA is recognized by a nuclear protein known as DiGeorge Syndrome Critical Region 8 (DGCR8 or "Pasha" in invertebrates), named for its association with DiGeorge Syndrome. DGCR8 associates with the enzyme Drosha, a protein that cuts RNA, to form the Microprocessor complex. In this complex, DGCR8 orients the catalytic RNase III domain of Drosha to liberate hairpins from pri-miRNAs by cleaving RNA about eleven nucleotides from the hairpin base (one helical dsRNA turn into the stem). The product resulting has a two-nucleotide overhang at its 3' end; it has 3' hydroxyl and 5' phosphate groups. It is often termed as a pre-miRNA (precursor-miRNA). Sequence motifs downstream of the pre-miRNA that are important for efficient processing have been identified.

Pre-miRNAs that are spliced directly out of introns, bypassing the Microprocessor complex, are known as "mirtrons." Mirtrons have been found in *Drosophila*, *C. elegans*, and mammals.

As many as 16% of pre-miRNAs may be altered through nuclear RNA editing. Most commonly, enzymes known as adenosine deaminases acting on RNA (ADARs) catalyze adenosine to inosine (A to I) transitions. RNA editing can halt nuclear processing (for example, of pri-miR-142, leading to degradation by the ribonuclease Tudor-SN) and alter downstream processes including cytoplasmic miRNA processing and target specificity (e.g., by changing the seed region of miR-376 in the central nervous system).

Nuclear Export

Pre-miRNA hairpins are exported from the nucleus in a process involving the nucleocytoplasmic shuttler Exportin-5. This protein, a member of the karyopherin family, recognizes a two-nucleotide overhang left by the RNase III enzyme Drosha at the 3' end of the pre-miRNA hairpin. Exportin-5-mediated transport to the cytoplasm is energy-dependent, using guanosine triphosphate (GTP) bound to the Ran protein.

Cytoplasmic Processing

In the cytoplasm, the pre-miRNA hairpin is cleaved by the RNase III enzyme Dicer. This endoribonuclease interacts with 5' and 3' ends of the hairpin and cuts away the loop joining the 3' and 5' arms, yielding an imperfect miRNA:miRNA* duplex about 22 nucleotides in length. Overall hairpin length and loop size influence the efficiency of Dicer processing. The imperfect nature of the miRNA:miRNA* pairing also affects cleavage. Some of the G-rich pre-miRNAs can potentially adopt the G-quadruplex structure as an alternative to the canonical stem-loop structure. For example, human pre-miRNA 92b adopts a G-quadruplex structure which is resistant to the Dicer mediated cleavage in the cytoplasm. Although either strand of the duplex may potentially act as a functional miRNA, only one strand is usually incorporated into the RNA-induced silencing complex (RISC) where the miRNA and its mRNA target interact.

While the majority of miRNAs are located within the cell, some miRNAs, commonly known as circulating miRNAs or extracellular miRNAs, have also been found in extracellular environment, including various biological fluids and cell culture media.

Biogenesis in Plants

miRNA biogenesis in plants differs from animal biogenesis mainly in the steps of nuclear processing and export. Instead of being cleaved by two different enzymes, once inside and once outside the nucleus, both cleavages of the plant miRNA are performed by a Dicer homolog, called Dicer-like1 (DL1). DL1 is expressed only in the nucleus of plant cells, which indicates that both reactions take place inside the nucleus. Before plant miRNA:miRNA* duplexes are transported out of the nucleus, its 3' overhangs are methylated by a RNA methyltransferaseprotein called Hua-Enhancer1 (HEN1). The duplex is then transported out of the nucleus to the cytoplasm by a protein called Hasty (HST), an Exportin 5 homolog, where they disassemble and the mature miRNA is incorporated into the RISC.

RNA-Induced Silencing Complex

The mature miRNA is part of an active RNA-induced silencing complex (RISC) containing Dicer and many associated proteins. RISC is also known as a microRNA ribonucleoprotein complex (miRNP); A RISC with incorporated miRNA is sometimes referred to as a "miRISC."

Dicer processing of the pre-miRNA is thought to be coupled with unwinding of the duplex. Generally, only one strand is incorporated into the miRISC, selected on the basis of its thermodynamic instability and weaker base-pairing on the 5' end relative to the other strand. The position of the stem-loop may also influence strand choice. The other strand, called the passenger strand due to its lower levels in the steady state, is denoted with an asterisk (*) and is normally degraded. In some cases, both strands of the duplex are viable and become functional miRNA that target different mRNA populations.

Members of the Argonaute (Ago) protein family are central to RISC function. Argonautes are needed for miRNA-induced silencing and contain two conserved RNA binding domains: a PAZ domain that can bind the single stranded 3' end of the mature miRNA and a PIWI domain that structurally resembles ribonuclease-H and functions to interact with the 5' end of the guide strand. They bind the mature miRNA and orient it for interaction with a target mRNA. Some argonautes, for example human Ago2, cleave target transcripts directly; argonautes may also recruit additional proteins to achieve translational repression. The human genome encodes eight argonaute proteins divided by sequence similarities into two families: AGO (with four members present in all mammalian cells and called E1F2C/hAgo in humans), and PIWI (found in the germline and hematopoietic stem cells).

Additional RISC components include TRBP [human immunodeficiency virus (HIV) transactivating response RNA (TAR) binding protein], PACT (protein activator of the interferon-induced protein kinase), the SMN complex, fragile X mental retardation protein (FMRP), Tudor staphylococcal nuclease-domain-containing protein (Tudor-SN), the putative DNA helicase MOV10, and the RNA recognition motif containing protein TNRC6B.

Mode of Silencing and Regulatory Loops

Gene silencing may occur either via mRNA degradation or preventing mRNA from being translated. For example, miR16 contains a sequence complementary to the AU-rich element found in the 3'UTR of many unstable mRNAs, such as TNF alpha or GM-CSF. It has been demonstrated that given complete complementarity between the miRNA and target mRNA sequence, Ago2 can cleave the mRNA and lead to direct mRNA degradation. In the absence of complementarity, silencing is achieved by preventing translation. The relation of miRNA and its target mRNA can be based on the simple negative regulation of a target mRNA, but it seems that a common scenario is the use of a "coherent feed-forward loop", "mutual negative feedback loop" (also termed double negative loop) and "positive feedback/feed-forward loop". Some miRNAs work as buffers of random gene expression changes arising due to stochastic events in transcription, translation and protein stability. Such regulation is typically achieved by the virtue of negative feedback loops or incoherent feed-forward loop uncoupling protein output from mRNA transcription.

Turnover

Turnover of mature miRNA is needed for rapid changes in miRNA expression profiles. During miRNA maturation in the cytoplasm, uptake by the Argonaute protein is thought to stabilize the guide strand, while the opposite (* or "passenger") strand is preferentially destroyed. In what has been called a "Use it or lose it" strategy, Argonaute may preferentially retain miRNAs with many targets over miRNAs with few or no targets, leading to degradation of the non-targeting molecules. Decay of mature miRNAs in *Caenorhabditis elegans* is mediated by the 5'-to-3' exoribonuclease XRN2, also known as Rat1p. In plants, SDN (small RNA degrading nuclease) family members degrade miRNAs in the opposite (3'-to-5') direction. Similar enzymes are encoded in animal genomes, but their roles have not been described.

Several miRNA modifications affect miRNA stability. As indicated by work in the model organism *Arabidopsis thaliana* (thale cress), mature plant miRNAs appear to be stabilized by the addition of methyl moieties at the 3' end. The 2'-O-conjugated methyl groups block the addition of uracil (U) residues by uridyltransferase enzymes, a modification that may be associated with miRNA degradation. However, uridylation may also protect some miRNAs; the consequences of this modification are incompletely understood. Uridylation of some animal miRNAs has been reported. Both plant and animal miRNAs may be altered by addition of adenine (A) residues to the 3' end of the miRNA. An extra A added to the end of mammalian miR-122, a liver-enriched miRNA important in hepatitis C, stabilizes the molecule and plant miRNAs ending with an adenine residue have slower decay rates.

Cellular Functions

The function of miRNAs appears to be in gene regulation. For that purpose, a miRNA is complementary to a part of one or more messenger RNAs (mRNAs). Animal miRNAs are usually complementary to a site in the 3' UTR whereas plant miRNAs are usually complementary to coding regions of mRNAs. Perfect or near perfect base pairing with the target RNA promotes cleavage of the RNA. This is the primary mode of plant miRNAs. In animals the match-ups are imperfect.

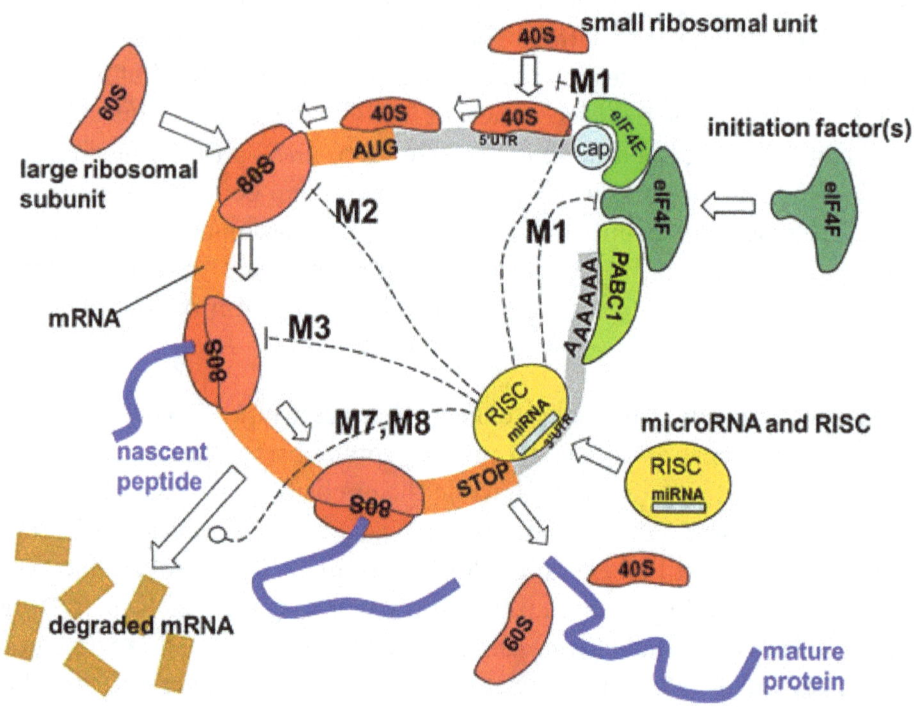

Fig: Interaction of microRNA with Protein Translation Process

Several translation repression mechanisms are shown: M1) on the initiation process, preventing assembling of the initiation complex or recruiting the 40S ribosomal subunit; M2) on the ribosome assembly; M3) on the translation process; M7, M8) on the degradation of mRNA. 40S and 60S are light and heavy components of the ribosome, 80S is the assembled ribosome bound to mRNA, eIF4F is a translation initiation factor, PABC1 is the Poly-A binding protein, and "cap" is the mRNA cap structure needed for mRNA circularization (which can be the normal m7G-cap or modified A-cap). The initiation of mRNA can proceed in a cap-independent manner, through recruiting 40S to IRES (Internal Ribosome Entry Site) located in 5'UTR region. The actual work of RNA silencing is performed by RISC in which the main catalytic subunit is one of the Argonaute proteins (AGO), and miRNA serves as a template for recognizing specific mRNA sequences.

For partially complementary microRNAs to recognise their targets, nucleotides 2–7 of the miRNA (its 'seed region') must be perfectly complementary. Animal miRNAs inhibit protein translation of the target mRNA (this is present but less common in plants). Partially complementary microRNAs can also speed up deadenylation, causing mRNAs to be degraded sooner. While degradation of miRNA-targeted mRNA is well documented, whether or not translational repression is accomplished through mRNA degradation, translational inhibition, or a combination of the two is hotly debated. Recent work on miR-430 in zebrafish, as well as on bantam-miRNA and miR-9 in *Drosophila* cultured cells, shows that translational repression is caused by the disruption of translation initiation, independent of mRNA deadenylation.

miRNAs occasionally also cause histone modification and DNA methylation of promoter sites, which affects the expression of target genes.

Nine mechanisms of miRNA action are described and assembled in a unified mathematical model:

- Cap-40S initiation inhibition;
- 60S Ribosomal unit joining inhibition;
- Elongation inhibition;
- Ribosome drop-off (premature termination);
- Co-translational nascent protein degradation;
- Sequestration in P-bodies;
- mRNA decay (destabilisation);
- mRNA cleavage; and
- Transcriptional inhibition through microRNA-mediated chromatin reorganization followed by gene silencing.

It is often impossible to discern these mechanisms using experimental data about stationary reaction rates. Nevertheless, they are differentiated in dynamics and have different *kinetic signatures*. Unlike plant microRNAs, the animal microRNAs target diverse genes.

However, genes involved in functions common to all cells, such as gene expression, have relatively fewer microRNA target sites and seem to be under selection to avoid targeting by microRNAs. There is a strong correlation between *ITPR* gene regulations and mir-92 and mir-19. dsRNA can also activate gene expression, a mechanism that has been termed "small RNA-induced gene activation" or RNAa. dsRNAs targeting gene promoters can induce potent transcriptional activation of associated genes. This was demonstrated in human cells using synthetic dsRNAs termed small activating RNAs (saRNAs), but has also been demonstrated for endogenous microRNA.

Interactions between microRNAs and complementary sequences on genes and even pseudogenes that share sequence homology are thought to be a back channel of communication regulating expression levels between paralogous genes (genes having a similar structure indicating divergence from a common ancestral gene). Given the name "competing endogenous RNAs" (ceRNAs), these microRNAs bind to "microRNA response elements" on genes and pseudogenes and may provide another explanation for the persistence of non-coding DNA.

miRNAs are also found as extracellular circulating miRNAs. Circulating miRNAs are released into body fluids including blood and cerebrospinal fluid and have the potential to be available as biomarkers in a number of diseases. Some researches show that mRNA cargo of exosomes may have a role in implantation, they can savage an adhesion between trophoblast and endometrium or support the adhesion by down regulating or up regulating expression of genes involved in adhesion/ invasion. Moreover, miRNA as miR-183/96/182 seems to play a key role in circadian rhythm.

Evolution

miRNAs are well conserved in both plants and animals, and are thought to be a vital and evolutionarily ancient component of gene regulation. While core components of the microRNA pathway are conserved between plants and animals, miRNA repertoires in the two kingdoms appear to have emerged independently with different primary modes of action.

microRNAs are useful phylogenetic markers because of their apparently low rate of evolution. microRNAs' origin as a regulatory mechanism developed from previous RNAi machinery that was initially used as a defense against exogenous genetic material such as viruses. Their origin may have permitted the development of morphological innovation, and by making gene expression more specific and 'fine-tunable', permitted the genesis of complex organs and perhaps, ultimately, complex life. Rapid bursts of morphological innovation are generally associated with a high rate of microRNA accumulation.

New microRNAs are created in multiple ways. Novel microRNAs can originate from the random formation of hairpins in "non-coding" sections of DNA (i.e., introns or intergene regions), but also by the duplication and modification of existing microRNAs. microRNAs can also form from inverted duplications of protein-coding sequences, which allows for the creation of a foldback hairpin structure.

The rate of evolution (i.e., nucleotide substitution) in recently originated microRNAs is comparable to that elsewhere in the non-coding DNA, implying evolution by neutral drift; however, older microRNAs have a much lower rate of change (often less than one substitution per hundred million years), suggesting that once a microRNA gains a function, it undergoes purifying selection. Individual regions within an miRNA gene face different evolutionary pressures, where regions that are vital for processing and function have higher levels of conservation. At this point, a microRNA is rarely lost from an animal's genome, although newer microRNAs (thus presumably non-functional) are frequently lost.

In *Arabidopsis thaliana*, the net flux of miRNA genes has been predicted to be between 1.2 and 3.3 genes per million years. This makes them a valuable phylogenetic marker, and they are being looked upon as a possible solution to outstanding phylogenetic problems such as the relationships of arthropods. On the other hand, in multiple cases microRNAs correlate poorly with phylogeny, and it is possible that their phylogenetic concordance largely reflects a limited sampling of microRNAs.

microRNAs feature in the genomes of most eukaryotic organisms, from the brown algae to the animals. However, the difference in how these microRNAs function and the way they are processed suggests that microRNAs arose independently in plants and animals.

Focusing on the animals, the genome of *Mnemiopsis leidyi* appears to lack recognizable microRNAs, as well as the nuclear proteins Drosha and Pasha, which are critical to canonical microRNA biogenesis. It is the only animal thus far reported to be missing Drosha. MicroRNAs play a vital role in the regulation of gene expression in all non-ctenophore animals investigated thus far except for *Trichoplax adhaerens*, the first known member of the phylum Placozoa.

Across all species, in excess of 5000 different miRNAs had been identified by March 2010. Whilst short RNA sequences (50 – hundreds of base pairs) of a broadly comparable function occur in bacteria, bacteria lack true microRNAs.

Experimental Detection and Manipulation

While researchers focused on miRNA expression in physiological and pathological processes, various technical variables related to microRNA isolation emerged. The stability of stored miRNA samples has been questioned. microRNAs degrade much more easily than mRNAs, partly due to their length, but also because of ubiquitously present RNases. This makes it necessary to cool samples on ice and use RNase-free equipment.

microRNA expression can be quantified in a two-step polymerase chain reaction process of modified RT-PCR followed by quantitative PCR. Variations of this method achieve absolute or relative quantification. miRNAs can also be hybridized to microarrays, slides or chips with probes to hundreds or thousands of miRNA targets, so that relative levels of miRNAs can be determined in different samples. microRNAs can be both discovered and profiled by high-throughput sequencing methods (microRNA sequencing).

The activity of an miRNA can be experimentally inhibited using a locked nucleic acid (LNA) oligo, a Morpholino oligo or a 2'-O-methyl RNA oligo. A specific miRNA can be silenced by a complementary antagomir. microRNA maturation can be inhibited at several points by steric-blocking oligos. The miRNA target site of an mRNA transcript can also be blocked by a steric-blocking oligo. For the "in situ" detection of miRNA, LNA or Morpholino probes can be used. The locked conformation of LNA results in enhanced hybridization properties and increases sensitivity and selectivity, making it ideal for detection of short miRNA.

High-throughput quantification of miRNAs is error prone, for the larger variance (compared to mRNAs) that comes with methodological problems. mRNA-expression is therefore often analyzed to check for miRNA-effects in their levels. Databases can be used to pair mRNA- and miRNA-data that predict miRNA-targets based on their base sequence. While this is usually done after miRNAs of interest have been detected (e. g. because of high expression levels), ideas for analysis tools that integrate mRNA- and miRNA-expression information have been proposed.

Human and Animal Diseases

Just as miRNA is involved in the normal functioning of eukaryotic cells, so has dysregulation of miRNA been associated with disease. A manually curated, publicly available database, miR2Disease, documents known relationships between miRNA dysregulation and human disease.

Inherited Diseases

- A mutation in the seed region of miR-96 causes hereditary progressive hearing loss.
- A mutation in the seed region of miR-184 causes hereditary keratoconus with anterior polar cataract.
- Deletion of the miR-17~92 cluster causes skeletal and growth defects.

Cancer

The first human disease known to be associated with miRNA deregulation was chronic lymphocytic leukemia. Many other miRNAs also have links with cancer and accordingly are sometimes referred to as "oncomirs". In malignant B cells miRNAs participate in pathways fundamental to B cell development like B-cell receptor (BCR) signalling, B-cell migration/adhesion, cell-cell interactions in immune niches and the production and class-switching of immunoglobulins. MiRNAs influence B cell maturation, generation of pre-, marginal zone, follicular, B1, plasma and memory B cells.

Another role for miRNA in cancers is to use their expression level for prognosis. In NSCLC samples, low miR-324a levels may serve as an indicator of poor survival. Either high miR-185 or low miR-133b levels may correlate with metastasis and poor survival in colorectal cancer.

Furthermore, specific miRNAs may be associated with certain histological subtypes of colorectal cancer. For instance, expression levels of miR-205 and miR-373 have been shown to be increased in mucinous colorectal cancers and mucin-producing Ulcerative Colitis-associated colon cancers, but not in sporadic colonic adenocarcinoma that lack mucinous components. In-vitro studies suggested that miR-205 and miR-373 may functionally induce different features of mucinous-associated neoplastic progression in intestinal epithelial cells.

Hepatocellular carcinoma cell proliferation may arise from miR-21 interaction with MAP2K3, a tumor repressor gene. Optimal treatment for cancer involves accurately identifying patients for risk-stratified therapy. Those with a rapid response to initial treatment may benefit from truncated treatment regimens, showing the value of accurate disease response measures. Cell-free circulating miRNAs (cimiRNAs) are highly stable in blood, are overexpressed in cancer and are quantifiable within the diagnostic laboratory. In classical Hodgkin lymphoma, plasma miR-21, miR-494, and miR-1973 are promising disease response biomarkers. Circulating miRNAs have the potential to assist clinical decision making and aid interpretation of positron emission tomography combined with computerized tomography. They can be performed at each consultation to assess disease response and detect relapse.

MicroRNAs have the potential to be used as tools or targets for treatment of different cancers. The specific microRNA, miR-506 has been found to work as a tumor antagonist in several studies. A significant number of cervical cancer samples were found to have downregulated expression of miR-506. Additionally, miR-506 works to promote apoptosis of cervical cancer cells, through its direct target hedgehog pathway transcription factor, Gli3.

DNA Repair and Cancer

Many miRNAs can directly target and inhibit cell cycle genes to control cell proliferation. A new strategy for tumor treatment is to inhibit tumor cell proliferation by repairing the defective miRNA pathway in tumors. Cancer is caused by the accumulation of mutations from either DNA damage or uncorrected errors in DNA replication. Defects in DNA repair cause the accumulation of mutations, which can lead to cancer. Several genes involved in DNA repair are regulated by microRNAs.

Germline mutations in DNA repair genes cause only 2–5% of colon cancer cases. However, altered expression of microRNAs, causing DNA repair deficiencies, are frequently associated with cancers and may be an important causal factor. Among 68 sporadic colon cancers with reduced expression of the DNA mismatch repair protein MLH1, most were found to be deficient due to epigenetic methylation of the CpG island of the MLH1 gene. However, up to 15% of MLH1-deficiencies in sporadic colon cancers appeared to be due to over-expression of the microRNA miR-155, which represses MLH1 expression.

In 29–66% of glioblastomas, DNA repair is deficient due to epigenetic methylation of the MGMT gene, which reduces protein expression of MGMT. However, for 28% of glioblastomas, the MGMT protein is deficient, but the MGMT promoter is not methylated. In glioblastomas without methylated MGMT promoters, the level of microRNA miR-181d is inversely correlated with protein expression of MGMT and the direct target of miR-181d is the MGMT mRNA 3'UTR (the three prime untranslated region of MGMT mRNA). Thus, in 28% of glioblastomas, increased expression of miR-181d and reduced expression of DNA repair enzyme MGMT may be a causal factor.

HMGA proteins (HMGA1a, HMGA1b and HMGA2) are implicated in cancer, and expression of these proteins is regulated by microRNAs. HMGA expression is almost undetectable in differentiated adult tissues, but is elevated in many cancers. HMGA proteins are polypeptides of ~100 amino acid residues characterized by a modular sequence organization. These proteins have three highly positively charged regions, termed AT hooks, that bind the minor groove of AT-rich DNA stretches in specific regions of DNA. Human neoplasias, including thyroid, prostatic, cervical, colorectal, pancreatic and ovarian carcinomas, show a strong increase of HMGA1a and HMGA1b proteins. Transgenic mice with HMGA1 targeted to lymphoid cells develop aggressive lymphoma, showing that high HMGA1 expression is associated with cancers and that HMGA1 can act as an oncogene. HMGA2 protein specifically targets the promoter of ERCC1, thus reducing expression of this DNA repair gene. ERCC1 protein expression was deficient in 100% of 47 evaluated colon cancers (though the extent to which HGMA2 was involved is not known).

Single Nucleotide polymorphisms (SNPs) can alter the binding of miRNAs on 3'UTRs for example the case of hsa-mir181a and hsa-mir181b on the CDON tumor suppressor gene.

Heart Disease

The global role of miRNA function in the heart has been addressed by conditionally inhibiting miRNA maturation in the murine heart. This revealed that miRNAs play an essential role during its development. miRNA expression profiling studies demonstrate that expression levels of specific miRNAs change in diseased human hearts, pointing to their involvement in cardiomyopathies. Furthermore, animal studies on specific miRNAs identified distinct roles for miRNAs both during heart development and under pathological conditions, including the regulation of key factors important for cardiogenesis, the hypertrophic growth response and cardiac conductance. Another role for miRNA in cardiovascular diseases is to use their expression levels for diagnosis, prognosis or risk stratification. miRNA's in animal models have also been linked to cholesterol metabolism and regulation.

- **miRNA-712:** Murine microRNA-712 is a potential biomarker (i.e., predictor) for atherosclerosis, a cardiovascular disease of the arterial wall associated with lipid retention and inflammation. Non-laminar blood flow also correlates with development of atherosclerosis as mechanosenors of endothelial cells respond to the shear force of disturbed flow (d-flow). A number of pro-atherogenic genes including matrix metalloproteinases (MMPs) are upregulated by d-flow, mediating pro-inflammatory and pro-angiogenic signals. These findings were observed in ligated carotid arteries of mice to mimic the effects of d-flow. Within 24 hours, pre-existing immature miR-712 formed mature miR-712 suggesting that miR-712 is flow-sensitive. Coinciding with these results, miR-712 is also upregulated in endothelial cells exposed to naturally occurring d-flow in the greater curvature of the aortic arch.
- **Origin:** Pre-mRNA sequence of miR-712 is generated from the murine ribosomal RN45s gene at the internal transcribed spacer region 2 (ITS2). XRN1 is an exonuclease that degrades the ITS2 region during processing of RN45s. Reduction of XRN1 under d-flow conditions therefore leads to the accumulation of miR-712.
- **Mechanism:** MiR-712 targets tissue inhibitor of metalloproteinases 3 (TIMP3). TIMPs normally regulate activity of matrix metalloproteinases (MMPs) which degrade the extracellular matrix (ECM). Arterial ECM is mainly composed of collagen and elastin fibers, providing the structural support and recoil properties of arteries. These fibers play

a critical role in regulation of vascular inflammation and permeability, which are important in the development of atherosclerosis. Expressed by endothelial cells, TIMP3 is the only ECM-bound TIMP. A decrease in TIMP3 expression results in an increase of ECM degradation in the presence of d-flow. Consistent with these findings, inhibition of pre-miR712 increases expression of TIMP3 in cells, even when exposed to turbulent flow. TIMP3 also decreases the expression of TNFα (a pro-inflammatory regulator) during turbulent flow. Activity of TNFα in turbulent flow was measured by the expression of TNFα-converting enzyme (TACE) in blood. TNFα decreased if miR-712 was inhibited or TIMP3 overexpressed, suggesting that miR-712 and TIMP3 regulate TACE activity in turbulent flow conditions. Anti-miR-712 effectively suppresses d-flow-induced miR-712 expression and increases TIMP3 expression. Anti-miR-712 also inhibits vascular hyperpermeability, thereby significantly reducing atherosclerosis lesion development and immune cell infiltration.

- **Human Homolog MicroRNA-205:** The human homolog of miR-712 was found on the RN45s homolog gene, which maintains similar miRNAs to mice. MiR-205 of humans share similar sequences with miR-712 of mice and is conserved across most vertebrates. MiR-205 and miR-712 also share more than 50% of the cell signaling targets, including TIMP3. When tested, d-flow decreased the expression of XRN1 in humans as it did in mice endothelial cells, indicating a potentially common role of XRN1 in humans.

Kidney Disease

Targeted deletion of Dicer in the FoxD1-derived renal progenitor cells in a murine model resulted in a complex renal phenotype including expansion of nephron progenitors, fewer renin cells, smooth muscle arterioles, progressive mesangial loss and glomerular aneurysms. High throughput whole transcriptome profiling of the FoxD1-Dicer knockout mouse model revealed ectopic upregulation of pro-apoptotic gene, Bcl2L11 (Bim) and dysregulation of the p53 pathway with increase in p53 effector genes including Bax, Trp53inp1, Jun, Cdkn1a, Mmp2, and Arid3a. p53 protein levels remained unchanged, suggesting that FoxD1 stromal miRNAs directly repress p53-effector genes. Using a lineage tracing approach followed by Fluorescent-activated cell sorting, miRNA profiling of the FoxD1-derived cells not only comprehensively defined the transcriptional landscape of miRNAs that are critical for vascular development, but also identified key miRNAs that are likely to modulate the renal phenotype in its absence. These miRNAs include miRs–10a, 18a, 19b, 24, 30c, 92a, 106a, 130a, 152, 181a, 214, 222, 302a, 370, and 381 that regulate

Bcl2L11 (Bim) and miRs–15b, 18a, 21, 30c, 92a, 106a, 125b–5p, 145, 214, 222, 296–5p and 302a that regulate p53-effector genes. Consistent with the profiling results, ectopic apoptosis was observed in the cellular derivatives of the FoxD1 derived progenitor lineage and reiterates the importance of renal stromal miRNAs in cellular homeostasis.

Nervous System

MiRNAs are crucial for the healthy development and function of the nervous system. Previous studies demonstrate that miRNAs can regulate neuronal differentiation and maturation at various stages. MiRNAs also play important roles in synaptic development (such as dendritogenesis or spine morphogenesis) and synaptic plasticity (contributing to learning and memory). Elimination of miRNA formation in mice by experimental silencing of Dicer has led to pathological outcomes, such as reduced neuronal size, motor abnormalities (when silenced in striatal neurons), and neurodegeneration (when silenced in forebrain neurons). Altered miRNA expression has been found in neurodegenerative diseases (such as Alzheimer's disease, Parkinson's disease, and Huntington's disease) as well as many psychiatric disorders (including epilepsy, schizophrenia, major depression, bipolar disorder, and anxiety disorders).

- **Stroke:** According to the Center for Disease Control and Prevention, Stroke is one of the leading causes of death and long-term disability in America. 87% of the cases are ischemic strokes, which results from blockage in the artery of the brain that carries oxygen-rich blood. The obstruction of the blood flow means the brain cannot receive necessary nutrients, such as oxygen and glucose, and remove wastes, such as carbon dioxide. miRNAs plays a role in posttranslational gene silencing by targeting genes in the pathogenesis of cerebral ischemia, such as the inflammatory, angiogenesis, and apoptotic pathway.
- **Alcoholism:** The vital role of miRNAs in gene expression is significant to addiction, specifically alcoholism. Chronic alcohol abuse results in persistent changes in brain function mediated in part by alterations in gene expression. miRNA global regulation of many downstream genes deems significant regarding the reorganization or synaptic connections or long term neural adaptations involving the behavioral change from alcohol consumption to withdrawal and/or dependence. Up to 35 different miRNAs have been found to be altered in the alcoholic post-mortem brain, all of which target genes that include the regulation of the cell cycle, apoptosis, cell adhesion, nervous system development and cell signaling. Altered miRNA levels were found in the medial prefrontal cortex of alcohol-dependent mice, suggesting the role of miRNA in

orchestrating translational imbalances and the creation of differentially expressed proteins within an area of the brain where complex cognitive behavior and decision making likely originate. miRNAs can be either upregulated or downregulated in response to chronic alcohol use. miR-206 expression increased in the prefrontal cortex of alcohol-dependent rats, targeting the transcription factor brain-derived neurotrophic factor (BDNF) and ultimately reducing its expression. BDNF plays a critical role in the formation and maturation of new neurons and synapses, suggesting a possible implication in synapse growth/synaptic plasticity in alcohol abusers. miR-155, important in regulating alcohol-induced neuroinflammation responses, was found to be upregulated, suggesting the role of microglia and inflammatory cytokines in alcohol pathophysiology. Downregulation of miR-382 was found in the nucleus accumbens, a structure in the basal forebrain significant in regulating feelings of reward that power motivational habits. miR-382 is the target for the dopamine receptor D1 (DRD1), and its overexpression results in the upregulation of DRD1 and delta fosB, a transcription factor that activates a series of transcription events in the nucleus accumbens that ultimately result in addictive behaviors. Alternatively, overexpressing miR-382 resulted in attenuated drinking and the inhibition of DRD1 and delta fosB upregulation in rat models of alcoholism, demonstrating the possibility of using miRNA-targeted pharmaceuticals in treatments.

Obesity

miRNAs play crucial roles in the regulation of stem cell progenitors differentiating into adipocytes. Studies to determine what role pluripotent stem cells play in adipogenesis, were examined in the immortalized human bone marrow-derived stromal cell line hMSC-Tert20. Decreased expression of miR-155, miR-221, and miR-222, have been found during the adipogenic programming of both immortalized and primary hMSCs, suggesting that they act as negative regulators of differentiation. Conversely, ectopic expression of the miRNAs 155, 221, and 222 significantly inhibited adipogenesis and repressed induction of the master regulators PPARγ and CCAAT/enhancer-binding protein alpha (CEBPA). This paves the way for possible genetic obesity treatments.

Another class of miRNAs that regulate insulin resistance, obesity, and diabetes, is the let-7 family. Let-7 accumulates in human tissues during the course of aging. When let-7 was ectopically overexpressed to mimic accelerated aging, mice became insulin-resistant, and thus more prone to high fat diet-induced obesity and diabetes. In contrast when let-7 was inhibited by injections of let-7-specific antagomirs, mice become more insulin-sensitive and remarkably resistant to high fat diet-induced obesity and diabetes. Not only could let-7 inhibition prevent obesity and diabetes, it could also reverse and cure the condition. These experimental findings suggest that let-7 inhibition could represent a new therapy for obesity and type 2 diabetes.

Hemostasis

miRNAs also play crucial roles in the regulation of complex enzymatic cascades including the hemostatic blood coagulation system. Large scale studies of functional miRNA targeting have recently uncovered rationale therapeutic targets in the hemostatic system. They have been directly linked to Calcium homeostasis in the endoplasmic reticulum, which is critical in cell differentiation in early development.

Viruses

Viral microRNAs play an important role in the regulation of gene expression of viral and/or host genes to benefit the virus. Hence, miRNAs play a key role in host–virus interactions and pathogenesis of viral diseases. The expression of transcription activators by *human herpesvirus-6* DNA is believed to be regulated by viral miRNA.

Target Prediction

miRNAs can bind to target messenger RNA (mRNA) transcripts of protein-coding genes and negatively control their translation or cause mRNA degradation. It is of key importance to identify the miRNA targets accurately. A comparison of the predictive performance of eighteen in silico algorithms is available. Large scale studies of functional miRNA targeting suggest that many functional miRNAs can be missed by target prediction algorithms.

9.2.2 Small Interfering RNAs (siRNAs)

Small interfering RNA (siRNA), sometimes known as short interfering RNA or silencing RNA, is a class of double-stranded non-coding RNA molecules, typically 20–24 base pairs in length, similar to microRNA (miRNA), and operating within the RNA interference (RNAi) pathway. It interferes with the expression of specific genes with complementary nucleotide sequences by degrading messenger RNA (mRNA) after transcription, preventing translation. It was discovered in 1998 by Andrew Fire at the Carnegie Institution for Science in Washington, D.C. and Craig Mello at the University of Massachusetts in Worcester.

Structure

Naturally occurring siRNAs have a well-defined structure that is a short (usually 20 to 24-bp) double-stranded RNA (dsRNA) with phosphorylated 5' ends and hydroxylated 3' ends with two overhanging nucleotides. The Dicer enzyme catalyzes production of siRNAs from long dsRNAs and small hairpin RNAs. siRNAs can also be introduced into cells by transfection. Since in principle any gene can be knocked down by a synthetic siRNA with a complementary sequence, siRNAs are an important tool for validating gene function and drug targeting in the post-genomic era.

History

In 1998, Andrew Fire at Carnegie Institution for Science in Washington DC and Craig Mello at University of Massachusetts in Worcester discovered the RNAi mechanism while working on the gene expression in the nematode, *Caenorhabditis elegans*. They won the Nobel prize for their research with RNAi in 2006. siRNAs and their role in post-transcriptional gene silencing (PTGS) was discovered in plants by David Baulcombe's group at the Sainsbury Laboratory in Norwich, England and reported in *Science* in 1999. Thomas Tuschl and colleagues soon reported in *Nature* that synthetic siRNAs could induce RNAi in mammalian cells. In 2001, the expression of a specific gene was successfully silenced by introducing chemically synthesized siRNA into mammalian cells (Tuschl et al.) These discoveries led to a surge in interest in harnessing RNAi for biomedical research and drug development. Significant developments in siRNA therapies have been made with both organic (carbon based) and inorganic (non-carbon based) nanoparticles, which have been successful in drug delivery to the brain, offering promising methods to deliver therapeutics into human subjects. However, human applications of siRNA have had significant limitations to its success. One of these being off-targeting. There is also a possibility that these therapies can trigger innate immunity. Animal models have not been successful in accurately representing the extent of this response in humans. Hence, studying the effects of siRNA therapies has been a challenge.

In recent years, siRNA therapies have been approved and new methods have been established to overcome these challenges. There are approved therapies available for commercial use and several currently in the pipeline waiting to get approval.

Mechanism

The mechanism by which natural siRNA causes gene silencing through repression of translation occurs as follows:

- Long dsRNA (which can come from hairpin, complementary RNAs, and RNA-dependent RNA polymerases) is cleaved by an endo-ribonuclease called Dicer. Dicer cuts the long dsRNA to form short interfering RNA or siRNA; this is what enables the molecules to form the RNA-Induced Silencing Complex (RISC).
- Once siRNA enters the cell it gets incorporated into other proteins to form the RISC.
- Once the siRNA is part of the RISC complex, the siRNA is unwound to form single stranded siRNA.

- The strand that is thermodynamically less stable due to its base pairing at the 5′end is chosen to remain part of the RISC-complex
- The single stranded siRNA which is part of the RISC complex now can scan and find a complementary mRNA
- Once the single stranded siRNA (part of the RISC complex) binds to its target mRNA, it induces mRNA cleavage.
- The mRNA is now cut and recognized as abnormal by the cell. This causes degradation of the mRNA and in turn no translation of the mRNA into amino acids and then proteins. Thus silencing the gene that encodes that mRNA.

siRNA is also similar to miRNA, however, miRNAs are derived from shorter stemloop RNA products. miRNAs typically silence genes by repression of translation and have broader specificity of action, while siRNAs typically work with higher specificity by cleaving the mRNA before translation, with 100% complementarity.

RNAi Induction Using siRNAs or their Biosynthetic Precursors

Gene knockdown by transfection of exogenous siRNA is often unsatisfactory because the effect is only transient, especially in rapidly dividing cells. This may be overcome by creating an expression vector for the siRNA. The siRNA sequence is modified to introduce a short loop between the two strands. The resulting transcript is a short hairpin RNA (shRNA), which can be processed into a functional siRNA by Dicer in its usual fashion. Typical transcription cassettes use an RNA polymerase III promoter (e.g., U6 or H1) to direct the transcription of small nuclear RNAs (snRNAs) (U6 is involved in RNA splicing; H1 is the RNase component of human RNase P). It is theorized that the resulting siRNA transcript is then processed by Dicer.

The gene knockdown efficiency can also be improved by using cell squeezing. The activity of siRNAs in RNAi is largely dependent on its binding ability to the RNA-induced silencing complex (RISC). Binding of the duplex siRNA to RISC is followed by unwinding and cleavage of the sense strand with endonucleases. The remaining anti-sense strand-RISC complex can then bind to target mRNAs for initiating transcriptional silencing.

RNA Activation

It has been found that dsRNA can also activate gene expression, a mechanism that has been termed "small RNA-induced gene activation" or RNAa. It has been shown that dsRNAs targeting gene promoters induce potent transcriptional activation of associated genes. RNAa was demonstrated in human cells using synthetic dsRNAs, termed "small activating RNAs" (saRNAs). It is currently not known how conserved RNAa is in other organisms. One report in the *Aedes aegypti* mosquito has shown there is some evidence for RNAa and can be achieved by short or long dsRNAs targeting promoter regions.

Post-Transcriptional Gene Silencing

The siRNA-induced post transcriptional gene silencing is initiated by the assembly of the RNA-induced silencing complex (RISC). The complex silences certain gene expression by cleaving the mRNA molecules coding the target genes. To begin the process, one of the two siRNA strands, the guide strand (anti-sense strand), will be loaded into the RISC while the other strand, the passenger strand (sense strand), is degraded. Certain Dicer enzymes may be responsible for loading the guide strand into RISC.

Then, the siRNA scans for and directs RISC to perfectly complementary sequence on the mRNA molecules. The cleavage of the mRNA molecules is thought to be catalyzed by the Piwi domain of Argonaute proteins of the RISC. The mRNA molecule is then cut precisely by cleaving the phosphodiester bond between the target nucleotides which are paired to siRNA residues 10 and 11, counting from the 5'end. This cleavage results in mRNA fragments that are further degraded by cellular exonucleases. The 5' fragment is degraded from its 3' end by exosome while the 3' fragment is degraded from its 5' end by 5' -3' exoribonuclease 1(XRN1). Dissociation of the target mRNA strand from RISC after the cleavage allow more mRNA to be silenced. This dissociation process is likely to be promoted by extrinsic factors driven by ATP hydrolysis.

Sometimes cleavage of the target mRNA molecule does not occur. In some cases, the endonucleolytic cleavage of the phosphodiester backbone may be suppressed by mismatches of siRNA and target mRNA near the cleaving site. Other times, the Argonaute proteins of the RISC lack endonuclease activity even when the target mRNA and siRNA are perfectly paired. In such cases, gene expression will be silenced by an miRNA induced mechanism instead

Piwi-interacting RNAs are responsible for the silencing of transposons and are not siRNAs. PIWI-interacting RNAs (piRNAs) are a recently-discovered class of small non-coding RNAs (ncRNAs) with a length of 21-35 nucleotides. They play a role in gene expression regulation, transposon silencing, and viral infection inhibition. Once considered as "dark matter" of ncRNAs, piRNAs emerged as important players in multiple cellular functions in different organisms.

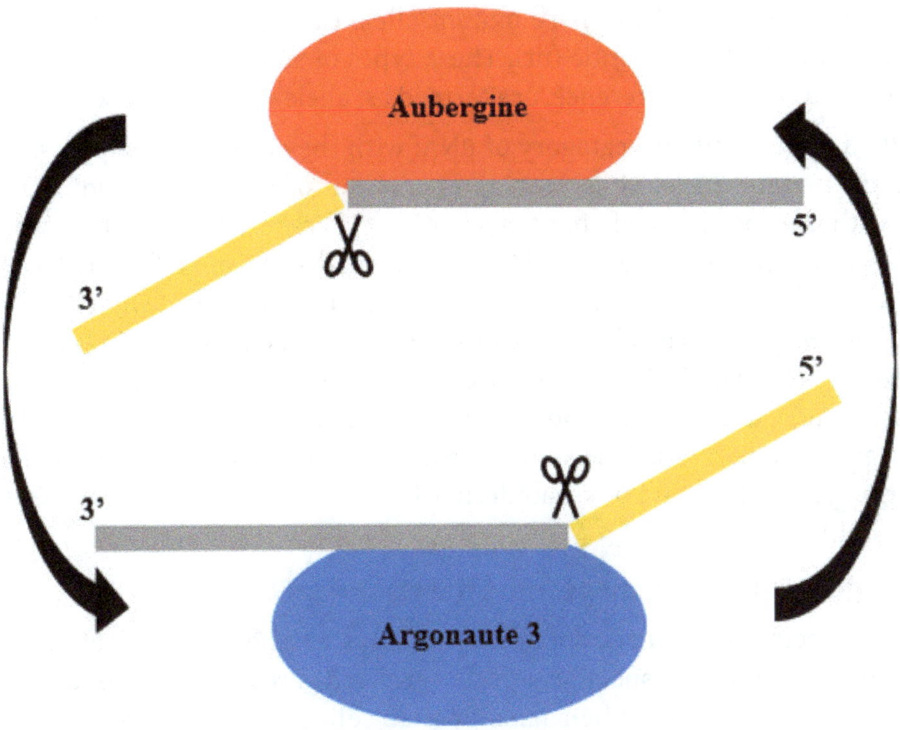

Fig: A Simplified Version of the Ping-Pong Method, Involving Proteins Aubergine (Aub) and Argonaute-3 (Ago3) Cleaving the 3' and 5' Ends of piRNA

Transcriptional Gene Silencing

Many model organism, such as plants (Arabidopsis thaliana), yeast (Saccharomyces cerevisiae), flies (Drosophila melanogaster) and worms (C. elegans), have been used to study small non coding RNA-directed Transcriptional gene silencing. In human cell, RNA-directed transcriptional gene silencing was observed a decade ago when exogenous siRNAs silenced a transgenic elongation factor 1 α promoter driving a Green Fluorescent Protein (GFP) reporter gene. The main mechanisms of transcriptional gene silencing (TGS) involving the RNAi machinery include DNA methylation, histone post-translational modifications, and subsequent chromatin remodeling around the target gene into a heterochromatic state. SiRNAs can be incorporated into a RNA-induced transcriptional silencing (RITS) complex. An active RITS complex will trigger the formation of heterochromatin around DNA matching the siRNA, effectively silencing the genes in that region of the DNA.

Applications: Allele-Specific Gene Silencing

One of the potent applications of siRNAs is the ability to distinguish the target versus non-target sequence with a single-nucleotide difference. This approach has been considered as therapeutically crucial for the silencing dominant gain-of-function (GOF) disorders,where mutant allele causing disease is differed from wt-allele by a single nucleotide (nt). These types of siRNAs with the capability to distinguish a single-nt difference, are termed as, allele-specific siRNAs.

ASP-RNAi is an innovative category of RNAi with the objective of suppressing the dominant mutant allele while sparing expression of the corresponding normal allele with the specificity of single-nucleotide differences between the two. ASP-siRNAs are potentially a novel and better remedial alternative for the treatment of autosomal dominant genetic disorders especially in cases where wild-type allele expression is crucial for organism survival such as Huntington disease (HD),DYT1 dystonia, Alzheimer's disease, Parkinson's disease (PD) , amyloid lateral sclerosis (ALS) and Machado–Joseph disease. Their therapeutic potential has also been assessed for various skin disorders like epidermolysis bullosa simplex, epidermolytic palmoplantar keratoderma (EPPK) and lattice corneal dystrophy type I (LCDI) .

Challenges: Avoiding Nonspecific Effects

RNAi intersects with a number of other pathways; as of 2010 it was not surprising that on occasion, nonspecific effects are triggered by the experimental introduction of an siRNA. When a mammalian cell encounters a double-stranded RNA such as an siRNA, it may mistake it as a viral by-product and mount an immune response.

Furthermore, because structurally related microRNAs modulate gene expression largely via incomplete complementarity base pair interactions with a target mRNA, the introduction of an siRNA may cause unintended off-targeting. Chemical modifications of siRNA may alter the thermodynamic properties that also result in a loss of single nucleotide specificity.

Innate Immunity

Introduction of too many siRNA can result in nonspecific events due to activation of innate immune responses. Most evidence to date suggests that this is probably due to activation of the dsRNA sensor PKR, although retinoic acid-inducible gene I (RIG-I) may also be involved.

The induction of cytokines via toll-like receptor 7 (TLR7) has also been described. Chemical modification of siRNA is employed to reduce in the activation of the innate immune response for gene function and therapeutic applications. One promising method of reducing the nonspecific effects is to convert the siRNA into a microRNA. MicroRNAs occur naturally, and by harnessing this endogenous pathway it should be possible to achieve similar gene knockdown at comparatively low concentrations of resulting siRNAs. This should minimize nonspecific effects.

Off-Targeting

Off-targeting is another challenge to the use of siRNAs as a gene knockdown tool. Here, genes with incomplete complementarity are inadvertently downregulated by the siRNA (in effect, the siRNA acts as a miRNA), leading to problems in data interpretation and potential toxicity. This, however, can be partly addressed by designing appropriate control experiments, and siRNA design algorithms are currently being developed to produce siRNAs free from off-targeting. Genome-wide expression analysis, e.g., by microarray technology, can then be used to verify this and further refine the algorithms. A 2006 paper from the laboratory of Dr. Khvorova implicates 6- or 7-basepair-long stretches from position 2 onward in the siRNA matching with 3'UTR regions in off-targeted genes. The tool of siRNA off-target predition is available at http://crdd.osdd.net/servers/aspsirna/asptar.php and published as ASPsiRNA resource.

Adaptive Immune Responses

Plain RNAs may be poor immunogens, but antibodies can easily be created against RNA-protein complexes. Many autoimmune diseases see these types of antibodies. There haven't yet been reports of antibodies against siRNA bound to proteins. Some methods for siRNA delivery adjoin polyethylene glycol (PEG) to the oligonucleotide reducing excretion and improving circulating half-life. However recently a large Phase III trial of PEGylated RNA aptamer against factor IX had to be discontinued by Regado Biosciences because of a severe anaphylactic reaction to the PEG part of the RNA. This reaction led to death in some cases and raises significant concerns about siRNA delivery when PEGylated oligonucleotides are involved.

Saturation of the RNAi Machinery

siRNAs transfection into cells typically lowers the expression of many genes, however, the upregulation of genes is also observed. The upregulation of gene expression can partially be explained by the predicted gene targets of endogenous miRNAs. Computational analyses of more than 150 siRNA transfection experiments support a model where exogenous siRNAs can saturate the endogenous RNAi machinery, resulting in the de-repression of endogenous miRNA-regulated genes. Thus, while siRNAs can produce unwanted off-target effects, i.e., unintended downregulation of mRNAs via a partial sequence match between the siRNA and target, the saturation of RNAi machinery is another distinct nonspecific effect, which involves the de-repression of miRNA-regulated genes and results in similar problems in data interpretation and potential toxicity.

Chemical Modification

siRNAs have been chemically modified to enhance their therapeutic properties, Short interfering RNA (siRNA) must be delivered to the site of action in the cells of target tissues in order for RNAi to fulfill its therapeutic promise. A detailed database of all such chemical modifications is manually curated as siRNAmod in scientific literature. Chemical modification of siRNA can also inadvertently result in loss of single-nucleotide specificity.

Therapeutic Applications and Challenges

Given the ability to knock down, in essence, any gene of interest, RNAi via siRNAs has generated a great deal of interest in both basic and applied biology. One of the biggest challenges to siRNA and RNAi based therapeutics is intracellular delivery. siRNA also has weak stability and pharmacokinetic behavior. Delivery of siRNA via nanoparticles has shown promise. siRNA oligos in vivo are vulnerable to degradation by plasma and tissue endonucleases and exonucleases and have shown only mild effectiveness in localized delivery sites, such as the human eye. Delivering pure DNA to target organisms is challenging because its large size and structure prevents it from diffusing readily across membranes. siRNA oligos circumvent this problem due to their small size of 21-23 oligos. This allows delivery via nano-scale delivery vehicles called nanovectors.

A good nanovector for siRNA delivery should protect siRNA from degradation, enrich siRNA in the target organ and facilitate the cellular uptake of siRNA. The three main groups of siRNA nanovectors are: lipid based, non-lipid organic-based, and inorganic. Lipid based nanovectors are excellent for delivering siRNA to solid tumors, but other cancers may require different non-lipid based organic nanovectors such as cyclodextrin based nanoparticles.

siRNAs delivered via lipid based nanoparticles have been shown to have therapeutic potential for central nervous system (CNS) disorders. Central nervous disorders are not uncommon, but the blood brain barrier (BBB) often blocks access of potential therapeutics to the brain. siRNAs that target and silence efflux proteins on the BBB surface have been shown to create an increase in BBB permeability. siRNA delivered via lipid based nanoparticles is able to cross the BBB completely.

A huge difficulty in siRNA delivery is the problem of off-targeting. Since genes are read in both directions, there exists a possibility that even if the intended antisense siRNA strand is read and knocks out the target mRNA, the sense siRNA strand may target another protein involved in another function.

Phase I results of the first two therapeutic RNAi trials (indicated for age-related macular degeneration, aka AMD) reported at the end of 2005 that siRNAs are well tolerated and have suitable pharmacokinetic properties.

In a phase 1 clinical trial, 41 patients with advanced cancer metastasised to liver were administered RNAi delivered through lipid nanoparticles. The RNAi targeted two genes encoding key proteins in the growth of the cancer cells, vascular endothelial growth factor, (VEGF), and kinesin spindle protein (KSP). The results showed clinical benefits, with the cancer either stabilized after six months, or regression of metastasis in some of the patients. Pharmacodynamic analysis of biopsy samples from the patients revealed the presence of the RNAi constructs in the samples, proving that the molecules reached the intended target.

Proof of concept trials have indicated that Ebola-targeted siRNAs may be effective as post-exposure prophylaxis in humans, with 100% of non-human primates surviving a lethal dose of Zaire Ebolavirus, the most lethal strain.

Legal Categorization and Legal Issues in a Near Future

Currently, SiRNA are currently chemically synthesized and so, are legally categorized inside EU and in USA as simple medicinal products. But as bioengineered siRNA (BERAs) are in development, these would be classified as biological medicinal products, at least in EU. The development of the BERAs technology raises the question of the categorization of drugs having the same mechanism of action but being produced chemically or biologically. This lack of consistency should be addressed.

Intracellular Delivery

There is great potential for RNA interference (RNAi) to be used therapeutically to reversibly silence any gene. For RNAi to realize its therapeutic potential, small interfering RNA (siRNA) must be delivered to the site of action in the cells of target tissues. But finding safe and efficient delivery mechanisms is a major obstacle to achieving the full potential of siRNA-based therapies. Unmodified siRNA is unstable in the bloodstream, has the potential to cause immunogenicity, and has difficulty readily navigating cell membranes. As a result, chemical alterations and/or delivery tools are needed to safely transfer siRNA to its site of action. There are three main techniques of delivery for siRNA that differ on efficiency and toxicity.

Transfection

In this technique siRNA first must be designed against the target gene. Once the siRNA is configured against the gene it has to be effectively delivered through a transfection protocol. Delivery is usually done by cationic liposomes, polymer nanoparticles, and lipid conjugation. This method is advantageous because it can deliver siRNA to most types of cells, has high efficiency and reproducibility, and is offered commercially. The most common commercial reagents for transfection of siRNA are Lipofectamine and Neon Transfection. However, it is not compatible with all cell types and has low in vivo efficiency.

Electroporation

Electrical pulses are also used to intracellularly deliver siRNA into cells. The cell membrane is made of phospholipids which makes it susceptible to an electric field. When quick but powerful electrical pulses are initiated the lipid molecules reorient themselves, while undergoing thermal phase transitions because of heating. This results in the making of hydrophilic pores and localized perturbations in the lipid bilayer cell membrane also causing a temporary loss of semipermeability. This allows for the escape of many intracellular contents, such as ions and metabolites as well as the simultaneous uptake of drugs, molecular probes, and nucleic acids. For cells that are difficult to transfect electroporation is advantageous however cell death is more probable under this technique. This method has been used to deliver siRNA targeting VEGF into the xenografted tumors in nude mice, which resulted in a significant suppression of tumor growth.

Viral-Mediated Delivery

The gene silencing effects of transfected designed siRNA are generally transient, but this difficulty can be overcome through an RNAi approach. Delivering this siRNA from DNA templates can be done through several recombinant viral vectors based on retrovirus, adeno-associated virus, adenovirus, and lentivirus. The latter is the most efficient virus that stably delivers siRNA to target cells as it can transduce nondividing cells as well as directly target the nucleus. These specific viral vectors have been synthesized to effectively facilitate siRNA that is not viable for transfection into cells. Another aspect is that in some cases synthetic viral vectors can integrate siRNA into the cell genome which allows for stable expression of siRNA and long-term gene knockdown. This technique is advantageous because it is in vivo and effective for difficult to transfect cell. However problems arise because it can trigger antiviral responses in some cell types leading to mutagenic and immunogenic effects. This method has potential use in gene silencing of the central nervous system for the treatment of Huntington's disease.

Therapies

A decade after the discovery of RNAi mechanism in 1993, the pharmaceutical sector heavily invested in the research and development of siRNA therapy. There are several advantages that this therapy has over small molecules and antibodies. It can be administered quarterly or every six months. Another advantage is that, unlike small molecule and monoclonal antibodies that need to recognize specific conformation of a protein, siRNA functions by Watson-Crick basepairing with mRNA. Therefore, any target molecule that needs to be treated with high affinity and specificity can be selected if the right nucleotide sequence is available. One of the biggest challenges researchers needed to overcome was the identification and establishment of a delivery system through which the therapies would enter the body. And that the immune system often mistakes the RNAi therapies as remnants of infectious agents, which can trigger an immune response. Animal models did not accurately represent the degree of immune response that was seen in humans and despite the promise in the treatment investors divested away from RNAi.

However, there were a few companies that continued with the development of RNAi therapy for humans. Alnylam Pharmaceuticals, Sirna Therapeutics and Dicerna Pharmaceuticals are few of the companies still working on bringing RNAi therapies to market. It was learned that almost all siRNA therapies administered in the bloodstream accumulated in the liver. That is why most of the early drug targets were diseases that affected the liver. Repeated developmental work also shed light on improving the chemical composition of the RNA molecule to reduce the immune response, subsequently causing little to no side effects.

Alnylam Pharmaceuticals

In 2018, Alnylam Pharmaceuticals became the first company to have a siRNA therapy approved by the FDA. Onpattro (patisiran) was approved for the treatment of polyneuropathy of hereditary transthyretin-mediated (hATTR) amyloidosis in adults. hATTR is a rare, progressively debilitating condition. During hATTR amyloidosis, misfolded transthyretin (TTR) protein is deposited in the extracellular space. Under typical folding conditions, TTR tetramers are made up of four monomers.

Hereditary ATTR amyloidosis is caused by a fault or mutation in the transthyretin (TTR) gene which is inherited. Changing just one amino-acid changes the tetrameric transthyretin proteins, resulting in unstable tetrameric transthyretin protein that aggregates in monomers and forms insoluble extracellular amyloid deposits. Amyloid buildup in various organ systems causes cardiomyopathy, polyneuropathy, gastrointestinal dysfunction. It affects 50,000 people worldwide. To deliver the drug directly to the liver, siRNA is encased in a lipid nanoparticle. The siRNA molecule halts the production of amyloid proteins by interfering with the RNA production of abnormal TTR proteins. This prevents the accumulation of these proteins in different organs of the body and helps the patients manage this disease.

Traditionally, liver transplantation has been the standard treatment for hereditary transthyretin amyloidosis, however its effectiveness may be limited by the persistent deposition of wild-type transthyretin amyloid after transplantation. There are also small molecule medications that provide temporary relief. Before Onpattro was released, the treatment options for hATTR were limited. After the approval of Onpattro, FDA awarded Alnylam with the Breakthrough Therapy Designation, which is given to drugs that are intended to treat a serious condition and are a substantial improvement over any available therapy. It was also awarded Orphan Drug Designations given to those treatments that are intended to safely treat conditions affecting less than 200,000 people.

Along with Onpattro, another RNA interference therapeutic drug has also been discovered (Partisiran) which has property of inhibiting hepatic synthesis of transthyretin. Target messenger RNA (mRNA) is cleaved as a result by tiny interfering RNAs coupled to the RNA-induced silencing complex. Patisiran, an investigational RNAi therapeutic drug, uses this process to decrease the production of mutant and wild-type transthyretin by cleaving on 3-untranslated region of transthyretin mRNA.

In 2019, FDA approved the second RNAi therapy, Givlaari (givosiran) used to treat acute hepatic porphyria (AHP). The disease is caused due to the accumulation of toxic porphobilinogen (PBG) molecules which are formed during the production of heme. These molecules accumulate in different organs and this can lead to the symptoms or attacks of AHP.

Givlaari is an siRNA drug that downregulates the expression of aminolevulinic acid synthase 1 (ALAS1), a liver enzyme involved in an early step in heme production. The downregulation of ALAS1 lowers the levels of neurotoxic intermediates that cause AHP symptoms.

Years of research has led to a greater understanding of siRNA therapies beyond those affecting the liver. As of 2019, Alnylam Pharmaceuticals was involved in therapies that may treat amyloidosis and CNS disorders like Huntington's disease and Alzheimer's disease. They have also partnered with Regeneron Pharmaceuticals to develop therapies for CNS, eye and liver diseases.

As of 2020, ONPATTRO and GIVLAARI, were available for commercial application, and two siRNAs, lumasiran (ALN-GO1) and inclisiran, have been submitted for new drug application to the FDA. Several siRNAs are undergoing phase 3 clinical studies, and more candidates are in the early developmental stage. In 2020, Alnylam and Vir pharmaceuticals announced a partnership and have started working on a RNAi therapy that would treat severe cases of COVID-19.

Other companies that have had success in developing a pipeline of siRNA therapies are Dicerna Pharmaceuticals, partnered Eli Lilly and Company and Arrowhead Pharmaceuticals partnered with Johnson and Johnson. Several other big pharmaceutical companies such as Amgen and AstraZeneca have also invested heavily in siRNA therapies as they see the potential success of this area of biological drugs.

9.3 RNA-INDUCED SILENCING COMPLEX (RISC) AND ITS FUNCTION

A critical component of RNAi is the RNA-induced silencing complex (RISC), which includes an Argonaute protein that plays a pivotal role in recognizing and binding to target mRNAs. The Argonaute protein within the RISC complex serves as a guide, pairing with the siRNA or miRNA to bind to complementary mRNA sequences.

Once bound, Argonaute cleaves or represses the mRNA, preventing it from being translated into a protein. The specificity of RISC for mRNA targets is determined by the small RNA it carries, making the complex highly versatile in regulating gene expression. By silencing specific mRNA transcripts, RISC effectively controls the production of particular proteins, thereby fine-tuning cellular responses to environmental signals or stress.

9.4 BIOLOGICAL ROLES OF RNA INTERFERENCE

RNA interference is not only a regulatory mechanism for gene expression but also plays various roles in cellular protection, development, and the maintenance of genomic integrity.

9.4.1 Defense against Viral Infections

RNAi serves as a critical defense mechanism against viral infections, especially in plants and invertebrates, where it acts as an antiviral immune response. When viruses infect cells, they often introduce double-stranded RNA, which is recognized by the cell's RNAi machinery.

This dsRNA is cleaved by Dicer into siRNAs, which are then used by the RISC complex to degrade viral RNA, preventing replication. This antiviral function of RNAi helps protect the cell from pathogen invasion and highlights RNAi's role in immune responses.

9.4.2 Regulation of Development and Differentiation

RNAi plays a key role in the regulation of genes involved in development and cell differentiation. By selectively silencing certain genes, RNAi controls developmental pathways and ensures that cells acquire specific functions. miRNAs, in particular, are involved in this process, as they regulate genes necessary for the timing of developmental stages and tissue-specific differentiation.

Abnormalities in miRNA function can lead to developmental defects, emphasizing RNAi's significance in orchestrating complex developmental processes.

9.4.3 Maintenance of Genomic Stability

Another important role of RNAi is maintaining genomic stability by silencing transposable elements and repetitive sequences. Transposable elements are segments of DNA that can move around within the genome, potentially causing mutations and genomic instability. siRNAs derived from these elements target and silence them, preventing their movement and ensuring the genome's integrity.

This function is particularly vital in germ cells, where protecting the integrity of genetic information is crucial for reproduction and heredity.

9.5 RNA INTERFERENCE AND DISEASE

Disruptions in RNAi processes are associated with various diseases, including cancer, neurodegenerative disorders, and viral infections. Because RNAi is essential for regulating gene expression, any dysregulation can lead to abnormal cellular behavior.

9.5.1 Cancer and Dysregulated miRNAs

In cancer, specific miRNAs known as oncomiRs can promote tumorigenesis by silencing tumor suppressor genes or activating oncogenes. These dysregulated miRNAs lead to abnormal cell proliferation, evasion of apoptosis, and metastasis. Understanding the role of miRNAs in cancer has led to potential therapeutic strategies that involve restoring or inhibiting specific miRNAs to treat cancer. For example, miRNA-based therapies are being explored to target and silence oncogenic pathways in cancer cells, offering new avenues for treatment.

9.5.2 Neurodegenerative Disorders and RNAi

Neurodegenerative diseases such as Alzheimer's and Parkinson's are associated with dysregulated RNAi pathways. In these conditions, the accumulation of toxic proteins or impaired degradation processes disrupt normal cellular function. Research into RNAi-based therapies aims to silence genes associated with protein aggregation, thereby reducing the harmful effects on neurons. Such approaches hold promise for developing treatments for conditions that currently have limited therapeutic options.

9.5.3 Viral Infections and RNAi-based Therapeutics

RNAi has been explored as a therapeutic approach against viral infections, including HIV, hepatitis, and SARS-CoV-2. By targeting viral RNA, RNAi-based therapies can prevent the virus from replicating within host cells. Although challenges exist, such as efficient delivery and avoidance of immune reactions, RNAi-based antiviral treatments are a promising avenue for combating viral diseases.

9.6 RNA INTERFERENCE AS A RESEARCH AND THERAPEUTIC TOOL

The discovery of RNAi has transformed genetic research, providing a method to silence genes with high specificity. This technique, known as "gene knockdown," allows researchers to study gene function by observing the effects of gene silencing.

9.6.1 RNAi in Genetic Research

RNAi is widely used in research to understand gene function by selectively silencing specific genes in model organisms such as fruit flies, worms, and mice. By observing the phenotypic changes that result from gene knockdown, researchers gain insight into the gene's role in development, metabolism, or disease. RNAi-based screens have enabled high-throughput genetic analysis, accelerating the discovery of genes involved in complex biological processes.

9.6.2 Therapeutic Potential of RNAi

The therapeutic potential of RNAi has led to the development of RNAi-based drugs targeting genetic and infectious diseases. For example, siRNA-based drugs have been approved for treating rare genetic disorders like hereditary transthyretin amyloidosis (hATTR). These drugs work by silencing the faulty gene responsible for the disorder, thereby reducing disease symptoms. RNAi-based therapeutics hold promise for a wide range of diseases, but challenges remain, including efficient delivery and avoiding unintended immune responses.

9.7 CHALLENGES AND FUTURE DIRECTIONS

Despite its potential, RNAi-based therapies face challenges, primarily in delivery and off-target effects. Delivering RNAi molecules into cells without degradation is difficult, as they are susceptible to degradation by cellular enzymes. Researchers are exploring delivery methods such as lipid nanoparticles and viral vectors to improve stability and uptake. Furthermore, off-target effects, where RNAi molecules silence unintended genes, can cause side effects. Advances in targeting and delivery methods will be crucial to realizing the full therapeutic potential of RNAi.

9.8 CONCLUSION

RNA interference is a powerful and versatile mechanism that regulates gene expression, protects the genome from viral and transposon activity, and maintains cellular function. It plays crucial roles in development, immune response, and disease prevention. The discovery of RNAi has revolutionized genetic research and holds immense potential for treating genetic and viral diseases. As research progresses, overcoming the challenges associated with RNAi delivery and specificity will bring us closer to harnessing this natural process for therapeutic benefit. The study and application of RNAi promise to expand our understanding of gene regulation and pave the way for innovative medical treatments.

PART 2:
FOUNDATIONS OF GENOMICS

CHAPTER 10

FUNDAMENTALS OF WHOLE-GENOME SEQUENCING

10.1 INTRODUCTION TO WHOLE-GENOME SEQUENCING

Whole-genome sequencing (WGS) is a comprehensive method used to determine the complete DNA sequence of an organism's genome. Unlike targeted sequencing techniques that focus on specific genes or regions, WGS provides a complete map of all the genes and non-coding regions within a genome. This powerful technology has revolutionized genomics by allowing scientists to explore genetic variation across individuals and species, understand disease mechanisms, and identify potential targets for therapeutic interventions. The availability of WGS has played a key role in advancing personalized medicine, evolutionary biology, and agricultural biotechnology.

10.2 THE PROCESS OF WHOLE-GENOME SEQUENCING

Whole-genome sequencing involves several steps, from DNA extraction to data analysis. The basic steps in WGS are:

10.2.1 DNA Extraction and Preparation

The first step in WGS is the extraction of DNA from a biological sample. The quality and quantity of the extracted DNA are crucial for successful sequencing. Common sources of DNA include blood, saliva, tissue biopsies, or even environmental samples. The DNA is then fragmented into smaller pieces to facilitate sequencing. These fragments are typically around 100 to 200 base pairs long. Fragmentation can be achieved through mechanical shearing or enzymatic digestion. Once fragmented, the DNA is prepared for sequencing by adding adapters to the ends of the fragments. These adapters are short, known sequences that allow the fragments to be recognized by the sequencing machine.

10.2.2 Sequencing

The sequencing step is the core of whole-genome sequencing. Several sequencing technologies exist, but the most widely used is next-generation sequencing (NGS), which enables high-throughput sequencing. In NGS, the DNA fragments are sequenced in parallel, generating millions of short reads. These reads are typically between 100 and 300 base pairs in length. The sequencing machine generates digital data representing the sequence of bases (adenine [A], cytosine [C], guanine [G], and thymine [T]) in the DNA fragments. Advances in NGS technology have made sequencing faster, cheaper, and more accurate, enabling WGS to become a routine procedure.

10.2.3 Assembly and Alignment

Once the sequencing data is generated, the next step is to assemble the short reads into longer contiguous sequences. This is done by comparing the overlapping regions of the reads and aligning them to generate a full genome sequence.

For many organisms, the genome is assembled by comparing the reads to a reference genome. In cases where a reference genome is not available, de novo assembly is used, which reconstructs the genome without relying on prior sequence information. This process can be computationally intensive, requiring powerful algorithms and substantial computing resources to produce accurate results.

10.2.4 Data Analysis

After sequencing and assembly, the resulting genome data undergoes extensive analysis. This step involves comparing the sequenced genome to known genetic markers and annotations to identify variations, such as single nucleotide polymorphisms (SNPs), insertions, deletions, and structural variants. Bioinformatics tools are used to annotate genes, predict the functional effects of mutations, and identify regions of interest for further study. Data analysis can also involve variant calling to determine differences between the sequenced genome and the reference genome, as well as identifying potential disease-causing mutations.

10.3 TECHNOLOGIES USED IN WHOLE-GENOME SEQUENCING

Whole-genome sequencing has evolved significantly since its inception, with new technologies continually improving the speed, accuracy, and affordability of sequencing. The major sequencing technologies used in WGS include:

10.3.1 Sanger Sequencing

Sanger sequencing, developed in the 1970s, was the first method used for DNA sequencing. Although it is no longer used for large-scale whole-genome sequencing due to its relatively low throughput, Sanger sequencing was instrumental in the completion of early genome projects, including the Human Genome Project. The method involves chain-termination, where DNA fragments are extended using modified nucleotides that terminate the chain. These fragments are then separated by size and detected to determine the DNA sequence. While highly accurate, Sanger sequencing is slower and more expensive than newer methods like NGS.

10.3.2 Next-Generation Sequencing (NGS)

NGS technologies, such as Illumina sequencing, Pacific Biosciences, and Oxford Nanopore Technologies, have revolutionized whole-genome sequencing by enabling parallel sequencing of millions of DNA fragments. These technologies allow for much faster, cheaper, and more scalable sequencing compared to Sanger sequencing. NGS generates large volumes of short reads, which are then assembled into complete genomes using sophisticated algorithms. Illumina sequencing, the most widely used NGS technology, works by sequencing DNA using fluorescently labeled nucleotides, while Pacific Biosciences and Oxford Nanopore provide long-read sequencing, which can help resolve complex genomic regions.

10.3.3 Long-Read Sequencing Technologies

Long-read sequencing technologies, such as Pacific Biosciences' SMRT (Single Molecule, Real-Time) sequencing and Oxford Nanopore sequencing, allow the generation of much longer reads than traditional NGS technologies. These longer reads are particularly useful for sequencing complex genomes, such as those with many repetitive sequences or structural variants, which can be difficult to resolve using short-read sequencing alone. Long-read sequencing helps to assemble more complete and accurate genomes and can provide valuable insights into gene structure and function.

10.4 APPLICATIONS OF WHOLE-GENOME SEQUENCING

Whole-genome sequencing has a wide range of applications in both research and clinical settings. Some of the most notable applications include:

10.4.1 Personalized Medicine

WGS has the potential to transform medicine by enabling personalized treatment plans based on an individual's genetic makeup. By identifying genetic variants associated with diseases or drug responses, doctors can tailor treatments to optimize effectiveness and minimize side effects. For example, certain genetic mutations may affect how patients metabolize medications, and WGS can help identify these individuals to ensure they receive the most suitable drugs and dosages. Personalized medicine also holds promise for the prevention and early detection of genetic diseases, such as cancer, by identifying predispositions before symptoms develop.

10.4.2 Disease Gene Identification

One of the primary uses of WGS is the identification of genes associated with rare and complex diseases. By sequencing the genomes of individuals affected by a particular disease, researchers can identify genetic mutations that contribute to the condition. WGS is particularly useful for identifying novel disease-causing genes and mutations that may not be detectable using traditional genetic tests. This has led to breakthroughs in understanding the genetic basis of disorders such as cystic fibrosis, Huntington's disease, and various types of cancer.

10.4.3 Evolutionary and Comparative Genomics

WGS also plays a vital role in evolutionary biology and comparative genomics. By sequencing the genomes of different species, researchers can compare genetic similarities and differences to better understand evolutionary relationships. For example, sequencing the genomes of extinct species like Neanderthals and comparing them to modern humans has provided valuable insights into human evolution. Comparative genomics also allows scientists to study the genetic basis of traits and adaptations in various organisms, including domesticated plants and animals.

10.4.4 Microbial Genomics

Whole-genome sequencing is widely used in microbial genomics to study bacteria, viruses, fungi, and other microorganisms. It enables the identification of microbial strains, tracking of outbreaks, and the study of antimicrobial resistance. Sequencing the genomes of pathogenic microbes can also help in developing vaccines and understanding the mechanisms of virulence and resistance.

10.5 CHALLENGES AND LIMITATIONS OF WHOLE-GENOME SEQUENCING

10.5.1 Cost

Although the cost of whole-genome sequencing has decreased significantly over the past decade, it is still relatively expensive compared to other genomic techniques. However, ongoing advancements in sequencing technology and computational methods are expected to further reduce costs, making WGS more accessible.

10.5.2 Data Analysis and Storage

Whole-genome sequencing generates enormous amounts of data, which presents significant challenges in terms of data analysis and storage. Interpreting the vast amount of genetic information requires advanced bioinformatics tools and computational power. Additionally, storing large genome datasets requires substantial storage capacity, which can be costly.

10.5.3 Interpretation of Variants

While WGS can identify a wide variety of genetic variants, interpreting the functional significance of these variants remains a challenge. Many variants identified through WGS are of unknown clinical significance, and understanding their impact on health requires further research. This is particularly true for variants found in non-coding regions of the genome, which were previously considered "junk DNA" but are now recognized as potentially important for gene regulation.

10.6 CONCLUSION

Whole-genome sequencing has emerged as a transformative technology with a wide range of applications across medicine, research, and biotechnology. By providing a complete map of an organism's genetic material, WGS enables the identification of genetic variations that contribute to disease, enhances our understanding of evolution, and supports personalized medicine.

While there are challenges related to cost, data analysis, and interpretation, ongoing technological advancements are expected to make WGS more accessible and effective. As WGS continues to evolve, it will undoubtedly lead to significant breakthroughs in our understanding of genetics and its role in health and disease.

CHAPTER 11

HOW THE HUMAN GENOME WAS SEQUENCED

11.1 INTRODUCTION TO THE HUMAN GENOME PROJECT

The Human Genome Project (HGP) was an international scientific research project with the goal of determining the base pairs that make up human DNA, and of identifying, mapping and sequencing all of the genes of the human genome from both a physical and a functional standpoint. It started in 1990 and was completed in 2003. It remains the world's largest collaborative biological project. Planning for the project began in 1984 by the US government, and it officially launched in 1990. It was declared complete on April 14, 2003, and included about 92% of the genome. Level "complete genome" was achieved in May 2021, with only 0.3% of the bases covered by potential issues. The final gapless assembly was finished in January 2022.

Funding came from the United States government through the National Institutes of Health (NIH) as well as numerous other groups from around the world. A parallel project was conducted outside the government by the Celera Corporation, or Celera Genomics, which was formally launched in 1998. Most of the government-sponsored sequencing was performed in twenty universities and research centres in the United States, the United Kingdom, Japan, France, Germany, and China, working in the International Human Genome Sequencing Consortium (IHGSC).

The Human Genome Project originally aimed to map the complete set of nucleotides contained in a human haploid reference genome, of which there are more than three billion. The *genome* of any given individual is unique; mapping the *human genome* involved sequencing samples collected from a small number of individuals and then assembling the sequenced fragments to get a complete sequence for each of the 23 human chromosome pairs (22 pairs of autosomes and a pair of sex chromosomes, known as allosomes). Therefore, the finished human genome is a mosaic, not representing any one individual. Much of the project's utility comes from the fact that the vast majority of the human genome is the same in all humans.

11.2 THE ORIGINS OF THE HUMAN GENOME PROJECT

The Human Genome Project was initiated by a global collaboration between researchers in the United States, Europe, and Asia. The motivation behind the project was to understand the genetic code of humans, which would provide insight into how genes influence health, disease, and development. The project was a collaborative effort of several institutions, with funding primarily provided by the U.S. government through the National Institutes of Health (NIH), as well as contributions from private organizations like the Wellcome Trust in the UK. The HGP was designed to address several scientific goals:

- To identify and map all of the genes in the human genome.
- To sequence the entire human genome.
- To store and freely disseminate the data to the global scientific community.
- To explore ethical, legal, and social issues related to human genomics.

11.3 STATE OF COMPLETION

Notably, the project was not able to sequence all of the DNA found in human cells; rather, the aim was to sequence only *euchromatic* regions of the nuclear genome, which make up 92.1% of the human genome. The remaining 7.9% exists in scattered *heterochromatic* regions such as those found in centromeres and telomeres. These regions by their nature are generally more difficult to sequence and so were not included as part of the project's original plans.

The Human Genome Project (HGP) was declared complete in April 2003. An initial rough draft of the human genome was available in June 2000 and by February 2001 a working draft had been completed and published followed by the final sequencing mapping of the human genome on April 14, 2003. Although this was reported to cover 99% of the euchromatic human genome with 99.99% accuracy, a major quality assessment of the human genome sequence was published on May 27, 2004, indicating over 92% of sampling exceeded 99.99% accuracy which was within the intended goal.

In March 2009, the Genome Reference Consortium (GRC) released a more accurate version of the human genome, but that still left more than 300 gaps, while 160 such gaps remained in 2015.

Though in May 2020, the GRC reported 79 "unresolved" gaps, accounting for as much as 5% of the human genome, months later, the application of new long-range sequencing techniques and a hydatidiform mole-derived cell line in which both copies of each chromosome are identical led to the first telomere-to-telomere, truly complete sequence of a human chromosome, the X chromosome. Similarly, an end-to-end complete sequence of human autosomal chromosome 8 followed several months later.

In 2021, it was reported that the Telomere-to-Telomere (T2T) consortium had filled in all of the gaps except five in repetitive regions of ribosomal DNA. Months later, those gaps had also been closed. The full sequence did not contain the Y chromosome, which causes the embryo to become male, being absent in the cell line that served as the source for the DNA analyzed. About 0.3% of the full sequence proved difficult to check for quality, and thus might have contained errors, which were being targeted for confirmation. In April 2022, the complete non-Y chromosome sequence was formally published, providing a view of much of the 8% of the genome left out by the HGP. In December, 2022, a preprint article claimed that the sequencing of the remaining missing regions of Y chromosome had been performed, thus completing the sequencing of all 24 human chromosomes. In August 2023 this preprint was finally published.

11.4 APPLICATIONS AND PROPOSED BENEFITS

The sequencing of the human genome holds benefits for many fields, from molecular medicine to human evolution. The Human Genome Project, through its sequencing of the DNA, can help researchers understand diseases including: genotyping of specific viruses to direct appropriate treatment; identification of mutations linked to different forms of cancer; the design of medication and more accurate prediction of their effects; advancement in forensic applied sciences; biofuels and other energy applications; agriculture, animal husbandry, bioprocessing; risk assessment; bioarcheology, anthropology and evolution. The sequence of the DNA is stored in databases available to anyone on the Internet. The U.S. National Center for Biotechnology Information (and sister organizations in Europe and Japan) house the gene sequence in a database known as GenBank, along with sequences of known and hypothetical genes and proteins. Other organizations, such as the UCSC Genome Browser at the University of California, Santa Cruz, and Ensembl present additional data and annotation and powerful tools for visualizing and searching it. Computer programs have been developed to analyze the data because the data itself is difficult to interpret without such programs. Generally speaking, advances in genome sequencing technology have followed Moore's Law, a concept from computer science which states that integrated circuits can increase in complexity at an exponential rate. This means that the speeds at which whole genomes can be sequenced can increase at a similar rate, as was seen during the development of the Human Genome Project.

11.5 TECHNIQUES AND ANALYSIS

The process of identifying the boundaries between genes and other features in a raw DNA sequence is called genome annotation and is in the domain of bioinformatics. While expert biologists make the best annotators, their work proceeds slowly, and computer programs are increasingly used to meet the high-throughput demands of genome sequencing projects. Beginning in 2008, a new technology known as RNA-seq was introduced that allowed scientists to directly sequence the messenger RNA in cells. This replaced previous methods of annotation, which relied on the inherent properties of the DNA sequence, with direct measurement, which was much more accurate. Today, annotation of the human genome and other genomes relies primarily on deep sequencing of the transcripts in every human tissue using RNA-seq. These experiments have revealed that over 90% of genes contain at least one and usually several alternative splice variants, in which the exons are combined in different ways to produce 2 or more gene products from the same locus.

The genome published by the HGP does not represent the sequence of every individual's genome. It is the combined mosaic of a small number of anonymous donors, of African, European and east Asian ancestry. The HGP genome is a scaffold for future work in identifying differences among individuals. Subsequent projects sequenced the genomes of multiple distinct ethnic groups, though as of 2019 there is still only one "reference genome".

11.5.1 Findings

Key findings of the draft (2001) and complete (2004) genome sequences include:

- There are approximately 22,300 protein-coding genes in human beings, the same range as in other mammals.
- The human genome has significantly more segmental duplications (nearly identical, repeated sections of DNA) than had been previously suspected.
- At the time when the draft sequence was published, fewer than 7% of protein families appeared to be vertebrate specific.

11.5.2 Accomplishments

The human genome has approximately 3.1 billion base pairs. The Human Genome Project was started in 1990 with the goal of sequencing and identifying all base pairs in the human genetic instruction set, finding the genetic roots of disease and then developing treatments. It is considered a megaproject.

The genome was broken into smaller pieces; approximately 150,000 base pairs in length. These pieces were then ligated into a type of vector known as "bacterial artificial chromosomes", or BACs, which are derived from bacterial chromosomes which have been genetically engineered. The vectors containing the genes can be inserted into bacteria where they are copied by the bacterial DNA replication machinery. Each of these pieces was then sequenced separately as a small "shotgun" project and then assembled. The larger, 150,000 base pairs go together to create chromosomes. This is known as the "hierarchical shotgun" approach, because the genome is first broken into relatively large chunks, which are then mapped to chromosomes before being selected for sequencing.

Funding came from the US government through the National Institutes of Health in the United States, and a UK charity organization, the Wellcome Trust, as well as numerous other groups from around the world. The funding supported a number of large sequencing centers including those at Whitehead Institute, the Wellcome Sanger Institute (then called The Sanger Centre) based at the Wellcome Genome Campus, Washington University in St. Louis, and Baylor College of Medicine.

The United Nations Educational, Scientific and Cultural Organization (UNESCO) served as an important channel for the involvement of developing countries in the Human Genome Project.

11.6 PUBLIC VERSUS PRIVATE APPROACHES

In 1998, a similar, privately funded quest was launched by the American researcher Craig Venter, and his firm Celera Genomics. Venter was a scientist at the NIH during the early 1990s when the project was initiated. The $300 million Celera effort was intended to proceed at a faster pace and at a fraction of the cost of the roughly $3 billion publicly funded project. While the Celera project focused its efforts on production sequencing and assembly of the human genome, the public HGP also funded mapping and sequencing of the worm, fly, and yeast genomes, funding of databases, development of new technologies, supporting bioinformatics and ethics programs, as well as polishing and assessment of the genome assembly. Both the Celera and public approaches spent roughly $250 million on the production sequencing effort. For sequence assembly, Celera made use of publicly available maps at GenBank, which Celera was capable of generating, but the availability of which was "beneficial" to the privately-funded project.

Celera used a technique called whole genome shotgun sequencing, employing pairwise end sequencing, which had been used to sequence bacterial genomes of up to six million base pairs in length, but not for anything nearly as large as the three billion base pair human genome.

Celera initially announced that it would seek patent protection on "only 200–300" genes, but later amended this to seeking "intellectual property protection" on "fully-characterized important structures" amounting to 100–300 targets. The firm eventually filed preliminary ("place-holder") patent applications on 6,500 whole or partial genes. Celera also promised to publish their findings in accordance with the terms of the 1996 "Bermuda Statement", by releasing new data annually (the HGP released its new data daily), although, unlike the publicly funded project, they would not permit free redistribution or scientific use of the data. The publicly funded competitors were compelled to release the first draft of the human genome before Celera for this reason. On July 7, 2000, the UCSC Genome Bioinformatics Group released a first working draft on the web. The scientific community downloaded about 500 GB of information from the UCSC genome server in the first 24 hours of free and unrestricted access.

In March 2000, President Clinton, along with Prime Minister Tony Blair in a dual statement, urged that all researchers who wished to research the sequence should have "unencumbered access" to the genome sequence. The statement sent Celera's stock plummeting and dragged down the biotechnology-heavy Nasdaq. The biotechnology sector lost about $50 billion in market capitalization in two days.

Although the working draft was announced in June 2000, it was not until February 2001 that Celera and the HGP scientists published details of their drafts. Special issues of *Nature* (which published the publicly funded project's scientific paper) described the methods used to produce the draft sequence and offered analysis of the sequence. These drafts covered about 83% of the genome (90% of the euchromatic regions with 150,000 gaps and the order and orientation of many segments not yet established). In February 2001, at the time of the joint publications, press releases announced that the project had been completed by both groups. Improved drafts were announced in 2003 and 2005, filling in to approximately 92% of the sequence currently.

11.7 GENOME DONORS

In the International Human Genome Sequencing Consortium (IHGSC) public-sector HGP, researchers collected blood (female) or sperm (male) samples from a large number of donors. Only a few of many collected samples were processed as DNA resources. Thus the donor identities were protected so neither donors nor scientists could know whose DNA was sequenced. DNA clones taken from many different libraries were used in the overall project, with most of those libraries being created by Pieter J. de Jong. Much of the sequence (>70%) of the reference genome produced by the public HGP came from a single anonymous male donor from Buffalo, New York, (code name RP11; the "RP" refers to Roswell Park Comprehensive Cancer Center).

HGP scientists used white blood cells from the blood of two male and two female donors (randomly selected from 20 of each) – each donor yielding a separate DNA library. One of these libraries (RP11) was used considerably more than others, because of quality considerations. One minor technical issue is that male samples contain just over half as much DNA from the sex chromosomes (one X chromosome and one Y chromosome) compared to female samples (which contain two X chromosomes). The other 22 chromosomes (the autosomes) are the same for both sexes.

Although the main sequencing phase of the HGP has been completed, studies of DNA variation continued in the International HapMap Project, whose goal was to identify patterns of single-nucleotide polymorphism (SNP) groups (called haplotypes, or "haps"). The DNA samples for the HapMap came from a total of 270 individuals; Yoruba people in Ibadan, Nigeria; Japanese people in Tokyo; Han Chinese in Beijing; and the French Centre d'Etude du Polymorphisme Humain (CEPH) resource, which consisted of residents of the United States having ancestry from Western and Northern Europe.

In the Celera Genomics private-sector project, DNA from five different individuals were used for sequencing. The lead scientist of Celera Genomics at that time, Craig Venter, later acknowledged (in a public letter to the journal *Science*) that his DNA was one of 21 samples in the pool, five of which were selected for use.

11.8 DEVELOPMENTS

With the sequence in hand, the next step was to identify the genetic variants that increase the risk for common diseases like cancer and diabetes. It is anticipated that detailed knowledge of the human genome will provide new avenues for advances in medicine and biotechnology. Clear practical results of the project emerged even before the work was finished. For example, a number of companies, such as Myriad Genetics, started offering easy ways to administer genetic tests that can show predisposition to a variety of illnesses, including breast cancer, hemostasis disorders, cystic fibrosis, liver diseases and many others. Also, the etiologies for cancers, Alzheimer's disease and other areas of clinical interest are considered likely to benefit from genome information and possibly may lead in the long term to significant advances in their management.

There are also many tangible benefits for biologists. For example, a researcher investigating a certain form of cancer may have narrowed down their search to a particular gene. By visiting the human genome database on the World Wide Web, this researcher can examine what other scientists have written about this gene, including (potentially) the three-dimensional structure of its product, its functions, its evolutionary relationships to other human genes, or to genes in mice, yeast, or fruit flies, possible detrimental mutations, interactions with other genes, body tissues in which this gene is activated, and diseases associated with this gene or other datatypes.

Further, a deeper understanding of the disease processes at the level of molecular biology may determine new therapeutic procedures. Given the established importance of DNA in molecular biology and its central role in determining the fundamental operation of cellular processes, it is likely that expanded knowledge in this area will facilitate medical advances in numerous areas of clinical interest that may not have been possible without them.

The analysis of similarities between DNA sequences from different organisms is also opening new avenues in the study of evolution. In many cases, evolutionary questions can now be framed in terms of molecular biology; indeed, many major evolutionary milestones (the emergence of the ribosome and organelles, the development of embryos with body plans, the vertebrate immune system) can be related to the molecular level. Many questions about the similarities and differences between humans and their closest relatives (the primates, and indeed the other mammals) are expected to be illuminated by the data in this project.

The project inspired and paved the way for genomic work in other fields, such as agriculture. For example, by studying the genetic composition of *Tritium aestivum*, the world's most commonly used bread wheat, great insight has been gained into the ways that domestication has impacted the evolution of the plant. It is being investigated which loci are most susceptible to manipulation, and how this plays out in evolutionary terms.

Genetic sequencing has allowed these questions to be addressed for the first time, as specific loci can be compared in wild and domesticated strains of the plant. This will allow for advances in the genetic modification in the future which could yield healthier and disease-resistant wheat crops, among other things.

11.9 ETHICAL, LEGAL, AND SOCIAL ISSUES

At the onset of the Human Genome Project, several ethical, legal, and social concerns were raised in regard to how increased knowledge of the human genome could be used to discriminate against people. One of the main concerns of most individuals was the fear that both employers and health insurance companies would refuse to hire individuals or refuse to provide insurance to people because of a health concern indicated by someone's genes.

In 1996, the United States passed the Health Insurance Portability and Accountability Act (HIPAA), which protects against the unauthorized and non-consensual release of individually identifiable health information to any entity not actively engaged in the provision of healthcare services to a patient.

Along with identifying all of the approximately 20,000–25,000 genes in the human genome (estimated at between 80,000 and 140,000 at the start of the project), the Human Genome Project also sought to address the ethical, legal, and social issues that were created by the onset of the project. For that, the Ethical, Legal, and Social Implications (ELSI) program was founded in 1990. Five percent of the annual budget was allocated to address the ELSI arising from the project. This budget started at approximately $1.57 million in the year 1990, but increased to approximately $18 million in the year 2014. Whilst the project may offer significant benefits to medicine and scientific research, some authors have emphasized the need to address the potential social consequences of mapping the human genome. Historian of science Hans-Jörg Rheinberger wrote that "the prospect of 'molecularizing' diseases and their possible cure will have a profound impact on what patients expect from medical help, and on a new generation of doctors' perception of illness."

In July 2024, an investigation by Undark Magazine and co-published with STAT News revealed for the first time several ethical lapses by the scientists spearheading the Human Genome Project. Chief among these was the use of roughly 75 percent of a single donor's DNA in the construction of the reference genome, despite informed consent forms, provided to each of the 20 anonymous donors participating, that indicated no more than 10 percent of any one donor's DNA would be used. About 10 percent of the reference genome belonged to one of the project's lead scientists, Pieter De Jong.

11.10 THE METHODOLOGY OF SEQUENCING THE HUMAN GENOME

Sequencing the human genome was an extraordinary technical challenge due to its size and complexity. Scientists needed to read and decode 3 billion base pairs of DNA from human chromosomes. The methodology used in the Human Genome Project can be broken down into key stages: DNA extraction, fragmentation, sequencing, assembly, and annotation.

11.10.1 DNA Extraction and Fragmentation

The first step in sequencing the human genome involved extracting DNA from human cells. The DNA was typically obtained from blood or tissue samples, which were processed to isolate the genetic material. Once the DNA was extracted, it had to be broken into smaller fragments. Since the entire human genome is too large to sequence in one go, scientists used enzymes or mechanical methods to fragment the DNA into smaller pieces, typically between 150 and 1,000 base pairs in length. This allowed for more manageable sequencing of smaller sections at a time.

11.10.2 Sequencing the Fragments

The next challenge was sequencing the millions of DNA fragments. Initially, researchers employed the Sanger sequencing method, which was widely used before the advent of next-generation sequencing technologies. Sanger sequencing uses a process known as chain termination to generate DNA fragments of varying lengths, which are then read by analyzing the color of fluorescently tagged nucleotides. These sequences were then reconstructed using computational tools.

However, the sheer volume of data required for the Human Genome Project demanded more efficient sequencing methods. The solution came in the form of high-throughput sequencing technologies, also known as next-generation sequencing (NGS). NGS methods, such as Illumina sequencing and Roche

454 sequencing, allowed scientists to sequence millions of DNA fragments simultaneously, significantly increasing the speed and cost-effectiveness of the project. These technologies generated vast amounts of data and reduced the time required for sequencing, accelerating the pace of the Human Genome Project.

11.10.3 Assembly of the Genome

Once the DNA fragments were sequenced, the next step was assembling the fragments back into a complete genome. This is a crucial and complex step because the genome is made up of millions of tiny pieces of DNA, each of which needs to be correctly aligned and ordered. Researchers used powerful computer algorithms to compare overlapping fragments and align them accurately, effectively piecing together the entire genome. Because human DNA is highly repetitive, this assembly process required sophisticated computational tools to distinguish between different repeats and accurately align them.

11.10.4 Annotation of the Genome

After sequencing and assembling the human genome, the next step was to annotate it—this means identifying the genes and other functional elements in the genome. Not all parts of the genome are genes; there are also non-coding regions, regulatory elements, and sequences whose function was not immediately clear. Researchers used comparative genomics and bioinformatics to identify and annotate the approximately 20,000-25,000 genes in the human genome, as well as regulatory regions involved in gene expression. The annotation process also involved identifying repetitive sequences and non-coding regions that might have previously been considered "junk DNA" but are now known to have important regulatory roles.

11.11 KEY MILESTONES IN THE SEQUENCING OF THE HUMAN GENOME

11.11.1 The Early Phase: 1990-1999

In the early phase of the HGP, the project was focused on the basic structure of the genome and the development of sequencing technologies. The first major milestone occurred in 1995 when scientists sequenced the genomes of several model organisms, such as the bacterium *Haemophilus influenzae*, marking the first fully sequenced organism. These achievements provided proof of principle for the larger task of sequencing the human genome.

11.11.2 The "Working Draft" of the Human Genome: 2001

In 2001, two separate international teams—one led by the Human Genome Project and the other by the private company Celera Genomics—announced a "working draft" of the human genome. This draft represented a significant achievement, with about 90% of the genome sequenced. However, many gaps and ambiguities remained, requiring further refinement and analysis.

11.11.3 Completion of the Human Genome Project: 2003

In 2003, the Human Genome Project officially announced the completion of the human genome sequence. While the initial goal was to finish by 2005, the project was completed two years ahead of schedule. By this point, 99% of the genome had been sequenced with a high degree of accuracy, and many gaps were filled in. The completion of the project was a monumental achievement, as it provided the first comprehensive map of the human genome.

11.12 CHALLENGES ENCOUNTERED DURING THE HUMAN GENOME PROJECT

11.12.1 Technological Limitations

In the early stages of the Human Genome Project, sequencing technology was relatively slow and expensive. Sanger sequencing, which was initially used, was a labor-intensive process, and the sheer size of the human genome made it difficult to scale. However, the development of next-generation sequencing (NGS) technologies, such as Illumina and 454 sequencing, transformed the project by allowing faster, cheaper, and more accurate sequencing.

11.12.2 Complex Genome Structure

The human genome is not just large but also complex. It contains repetitive sequences, structural variations, and complex gene regulation mechanisms. These complexities made assembling the genome a difficult task, as the repeated sequences could easily be misassembled or overlooked. Additionally, many regions of the genome are not encoded into proteins but serve regulatory or structural functions that were not fully understood at the time.

11.12.3 Ethical and Social Considerations

Sequencing the human genome raised a host of ethical and social issues, particularly regarding privacy, genetic testing, and potential discrimination. Concerns about how genetic information would be used—by employers, insurance companies, or governments—were central to discussions surrounding the Human Genome Project. In response, laws such as the Genetic Information Nondiscrimination Act (GINA) were enacted to protect individuals from genetic discrimination.

11.13 IMPACT OF SEQUENCING THE HUMAN GENOME

The completion of the Human Genome Project has had profound implications for many fields, including medicine, genetics, and evolutionary biology.

11.13.1 Advancements in Medicine

The knowledge gained from sequencing the human genome has paved the way for personalized medicine. By understanding the genetic basis of diseases, clinicians can now tailor treatments to individual genetic profiles. Genetic tests are increasingly used to identify predispositions to certain diseases, guide drug treatments, and develop gene therapies.

11.13.2 Insights into Human Evolution

Sequencing the human genome has also provided valuable insights into human evolution. By comparing the human genome to the genomes of other species, such as chimpanzees and Neanderthals, researchers have gained a better understanding of our evolutionary history and the genetic differences that make us uniquely human.

11.14 CONCLUSION

The sequencing of the human genome represents one of the greatest scientific achievements of modern times. Through international collaboration and the development of new sequencing technologies, scientists have unlocked the genetic blueprint of humanity. This knowledge has far-reaching implications for medicine, genetics, and our understanding of human evolution. As we continue to explore the human genome, the potential for new discoveries and advancements in healthcare is limitless.

PART 3:

ANALYZING GENOMES

CHAPTER 12
COMPARING AND ANALYZING GENOMES

The comparison and analysis of genomes have become crucial tools in modern genetics, providing insights into evolutionary biology, disease mechanisms, and the diversity of life. As sequencing technologies have advanced, researchers are now able to sequence entire genomes from various organisms and compare them to understand genetic similarities and differences. By comparing genomes, scientists can identify conserved genetic regions, track evolutionary changes, and gain a deeper understanding of the functions of specific genes. This process has vast applications in fields such as personalized medicine, agriculture, and anthropology.

12.1 WHAT IS GENOME COMPARISON?

Genome comparison involves the analysis of the sequence of DNA from different organisms to identify similarities and differences. These comparisons can be made between closely related species or between individuals within a single species. The primary goal of genome comparison is to understand the functional and evolutionary implications of these genetic differences. It also helps identify genes responsible for traits or diseases and gives clues about how genomes evolve over time.

The comparison of genomes typically involves the following key aspects:

- **Sequence Alignment**: This is the process of lining up two or more genome sequences to identify regions of similarity or difference.
- **Functional Annotation**: Identifying the functional elements within the genomes being compared, such as genes, regulatory regions, and other functional DNA sequences.
- **Phylogenetic Analysis**: Understanding the evolutionary relationships between different species based on genetic similarities and differences.

12.2 METHODS OF COMPARING GENOMES

Advancements in computational biology and sequencing technologies have enabled the development of various methods for comparing and analyzing genomes.

These methods allow for the identification of genetic similarities, differences, and evolutionary patterns across species or individuals. Some of the key methods include:

12.2.1 Sequence Alignment

Sequence alignment is the fundamental technique used to compare genomes. It involves aligning two or more sequences of DNA to identify matching regions and highlight areas of difference. There are two main types of sequence alignment:

- **Global Alignment**: This method compares entire genome sequences from start to finish. It is typically used when comparing two sequences that are similar in length and structure.
- **Local Alignment**: This method identifies regions of similarity within larger sequences, often used when comparing genomes that have different lengths or when focusing on specific regions of interest.

Multiple sequence alignment tools, such as ClustalW, MAFFT, and MUSCLE, are commonly used to align and compare large datasets of genomic sequences. These tools help identify conserved sequences that may be crucial for the organism's function and evolution.

12.2.2 SNP and Insertion/Deletion (Indel) Analysis

Single nucleotide polymorphisms (SNPs) and insertions/deletions (indels) are the most common types of genetic variation among genomes. SNPs are changes in a single base pair in the DNA sequence, while indels are variations caused by the insertion or deletion of one or more base pairs. By analyzing SNPs and indels across different genomes, scientists can pinpoint genetic variations that contribute to specific traits or diseases.

Next-generation sequencing (NGS) technologies have revolutionized the detection of SNPs and indels. These high-throughput methods allow for the sequencing of millions of DNA fragments simultaneously, enabling large-scale comparisons of genetic variation between individuals or species. Tools like GATK (Genome Analysis Toolkit) and Samtools are frequently used to identify and analyze these genetic variations.

12.2.3 Gene Family Comparison

Genomic comparisons also involve analyzing gene families, which are groups of related genes that share similar sequences and often perform similar functions. By comparing the gene families across different genomes, scientists can trace the evolution of specific gene families and understand how they have adapted to the needs of different organisms.

For example, the comparison of gene families between humans and other primates can provide insights into the genetic basis of human-specific traits such as language, cognition, and social behavior. Tools like OrthoMCL and InParanoid are used to identify and compare orthologous (derived from a common ancestor) and paralogous (resulting from gene duplication events) genes across different species.

12.2.4 Structural Genomics Comparison

Beyond the sequence itself, the structure of the genome also plays a critical role in its function. Structural genomics comparison involves analyzing the arrangement of genes, regulatory elements, and other genomic features, such as chromosomal inversions, translocations, and duplications. The comparison of the physical arrangement of genes across species can provide important clues about evolutionary processes and functional adaptations.

For instance, the human genome has a larger number of duplications compared to the genomes of other primates, which has contributed to our greater genetic diversity. Similarly, the comparison of genome structure between different species of plants or animals can reveal how these organisms have adapted to their environments over time.

12.2.5 Phylogenetic Analysis and Evolutionary Relationships

One of the primary uses of genome comparison is to explore the evolutionary relationships between different species. By comparing genomes, researchers can construct phylogenetic trees, which show the evolutionary connections between organisms. Phylogenetic analysis uses genetic data to determine how species are related and when they diverged from a common ancestor.

Genomic data is particularly useful for resolving phylogenetic relationships that were previously difficult to determine using morphological traits alone. For example, genomic comparisons have clarified the evolutionary relationship between humans and Neanderthals, showing that modern humans share a small but significant amount of DNA with Neanderthals, reflecting interbreeding between the two species.

12.3 APPLICATIONS OF GENOME COMPARISON

12.3.1 Personalized Medicine

The ability to compare the genomes of individuals has paved the way for personalized medicine, where treatments and therapies are tailored to an individual's genetic makeup. By identifying genetic variations associated with diseases, such as cancer or diabetes, researchers can develop targeted therapies that are more effective and have fewer side effects. Genome-wide association studies (GWAS) use large-scale genome comparisons to identify genetic risk factors for common diseases, enabling better predictions of disease susceptibility and treatment response.

12.3.2 Comparative and Evolutionary Genomics

Genome comparison is central to the field of evolutionary biology, as it allows scientists to trace the genetic changes that have occurred over time. By comparing the genomes of different species, researchers can identify conserved regions that are essential for basic biological functions and understand how genetic variation has contributed to species divergence. In addition to tracing evolutionary history, comparative genomics has been used to study speciation and adaptation. For example, comparing the genomes of domesticated plants and animals with their wild counterparts has provided insights into the genetic changes that occur during domestication.

12.3.3 Microbial Genomics and Antibiotic Resistance

Genome comparison is also crucial for studying microbes, including bacteria and viruses. By sequencing and comparing the genomes of different strains of bacteria, scientists can track the spread of diseases, identify antibiotic-resistant strains, and develop new treatments. The comparison of viral genomes has been instrumental in understanding the genetic variation of pathogens like the human immunodeficiency virus (HIV) and the SARS-CoV-2 virus, which causes COVID-19.

12.3.4 Agricultural Improvements

In agriculture, genome comparison is used to improve crop and livestock breeding. By comparing the genomes of different plant varieties or animal breeds, scientists can identify genes that confer desirable traits, such as disease resistance, drought tolerance, or higher yields. These genetic insights can be used to breed new varieties of crops or livestock that are better suited to specific environmental conditions.

12.4 CHALLENGES IN GENOME COMPARISON

12.4.1 Complexity of Genomes

Genomes are not uniform; they can contain highly repetitive regions, structural variations, and non-coding regions that are difficult to analyze. The presence of large, repetitive sequences in genomes, particularly in humans, complicates the alignment and comparison process. Additionally, structural variations, such as duplications and inversions, can introduce complications in genome assembly and comparison.

12.4.2 Data Interpretation

The sheer volume of data generated by sequencing technologies presents another challenge. Even after a genome has been sequenced, interpreting the results can be difficult. Not all genetic differences are functionally significant, and it can be challenging to determine which variations are associated with disease or other traits. Bioinformatics tools and algorithms are continually improving, but there is still much to learn about how genetic variations affect phenotypes.

12.5 CONCLUSION

Comparing and analyzing genomes is an essential aspect of modern genetics that has led to groundbreaking discoveries in evolution, medicine, and biodiversity. Through various methods such as sequence alignment, SNP analysis, gene family comparison, and phylogenetic analysis, scientists are uncovering the genetic underpinnings of life on Earth. The ability to compare genomes has wide-ranging applications, from personalized medicine to agriculture and evolutionary biology. Despite the challenges, ongoing advancements in sequencing technologies and computational tools promise to further enhance our understanding of genomes and their roles in health, disease, and evolution.

CHAPTER 13

FROM GENOME SEQUENCE TO GENE FUNCTION

13.1 INTRODUCTION: THE JOURNEY FROM GENOME TO FUNCTION

The sequencing of entire genomes has been a revolutionary step in biology, unlocking an immense amount of information about the genetic blueprint of organisms. However, while sequencing provides us with the raw genetic data, the next step is to translate this sequence into a deeper understanding of how genes function and influence biological processes. This transformation from the static sequence of nucleotides to dynamic, functional insights involves a complex interplay of genomic, transcriptomic, proteomic, and phenotypic data.

13.2 THE CHALLENGE OF GENE FUNCTION

The genome of an organism contains all the genetic information required for building and maintaining that organism. However, simply knowing the sequence of nucleotides in DNA does not reveal the full complexity of gene function. The challenge lies in deciphering how these sequences are read, how they interact with one another, and how their products—primarily proteins—affect cellular functions. In fact, many of the sequences in a genome do not directly encode proteins but instead have regulatory or structural roles that are equally critical for life. Moreover, genes rarely function in isolation. The regulation of gene expression, interactions between proteins, and the influence of environmental factors all contribute to the ultimate role of a gene in the organism's biology. Thus, understanding gene function is a multi-layered endeavor that extends far beyond reading the genome sequence itself.

13.3 THE ROLE OF GENE EXPRESSION IN FUNCTION

Gene expression refers to the process by which information from a gene is used to synthesize functional products, such as proteins or RNA molecules. This process includes transcription, in which an RNA copy of a gene's DNA sequence is made, and translation, in which the RNA is used as a template to synthesize proteins. Understanding gene function begins with understanding when and where genes are expressed, as well as the amount of expression.

13.3.1 Transcriptional Regulation

Before a gene is expressed as a protein, it must first be transcribed into messenger RNA (mRNA). The regulation of transcription is one of the key mechanisms controlling gene function. Transcription factors—proteins that bind to specific DNA sequences—either promote or inhibit the transcription of target genes. Additionally, epigenetic modifications, such as DNA methylation and histone modification, can influence transcription by altering chromatin structure and accessibility. For example, in multicellular organisms like humans, different cell types express different sets of genes. This differential gene expression underlies the diversity of cell types, from skin cells to neurons. Understanding how genes are regulated in different tissues and under different conditions is essential for deciphering gene function.

13.3.2 RNA Processing and Alternative Splicing

Once a gene is transcribed into mRNA, the primary transcript often undergoes various modifications before it is used to synthesize proteins. These modifications include the addition of a 5' cap and a 3' poly-A tail, as well as the splicing of introns (non-coding regions) and the joining of exons (coding regions). The manner in which a gene is spliced can produce different protein isoforms from the same gene, a phenomenon known as alternative splicing. Alternative splicing plays a crucial role in increasing the diversity of proteins produced by a genome. For instance, a single gene may be spliced in different ways in different tissues, leading to the production of distinct protein isoforms that have specific functions in different cellular environments. Understanding the mechanisms of RNA processing and splicing is therefore a key component in linking genome sequences to gene function.

13.4 TRANSLATING THE GENE: PROTEINS AND THEIR FUNCTIONS

Proteins are the functional products of most genes, and they carry out the vast majority of cellular functions. The synthesis of proteins from mRNA is called translation, and it occurs in the ribosomes, where tRNA molecules bring amino acids to form a polypeptide chain. This chain folds into a specific three-dimensional shape, which determines the protein's function.

13.4.1 Protein Structure and Function

The structure of a protein is intricately tied to its function. A protein's sequence of amino acids dictates its three-dimensional structure, which in turn determines how it interacts with other molecules in the cell. Proteins may function as enzymes, structural components, signaling molecules, or regulators of other genes. Understanding the relationship between a protein's structure and its function is key to understanding how genes influence biological processes. For example, mutations in the genes encoding enzymes can lead to diseases such as phenylketonuria (PKU), where a mutation in the phenylalanine hydroxylase gene results in the inability to metabolize the amino acid phenylalanine, leading to toxic accumulation. In such cases, understanding the function of the enzyme, its active sites, and how mutations affect its activity is essential for understanding the consequences of the genetic alteration.

13.4.2 Proteomics: Large-Scale Protein Analysis

Proteomics is the study of the entire complement of proteins produced by a genome. Unlike the genome, which is relatively constant, the proteome is highly dynamic and changes in response to cellular signals, environmental conditions, and genetic variations. High-throughput techniques such as mass spectrometry and protein microarrays are used to identify and quantify proteins across various conditions, providing a comprehensive view of gene function. Proteomics can reveal how proteins interact within the cell, which is essential for understanding cellular processes like signal transduction, metabolism, and gene regulation. By linking proteomic data to genomic information, scientists can build a more complete picture of how genes contribute to an organism's biology.

13.5 GENETIC VARIATION AND ITS IMPACT ON GENE FUNCTION

Genetic variation is a natural part of evolution, and understanding how variations in the genome affect gene function is a central aim of modern genetics. Variations can take many forms, including point mutations, insertions or deletions, and copy number variations. These genetic alterations can affect how genes are expressed, how their products function, or how they interact with other genes.

13.5.1 Single Nucleotide Polymorphisms (SNPs)

SNPs are the most common type of genetic variation and involve a change in a single nucleotide base in the genome. Some SNPs are silent and do not affect gene function, while others can alter protein structure, protein expression levels, or regulatory elements, thereby affecting gene function. Genome-wide association studies (GWAS) have been instrumental in linking specific SNPs to various diseases, traits, and conditions, revealing the functional consequences of these genetic variations. For instance, certain SNPs in the BRCA1 and BRCA2 genes are associated with an increased risk of breast cancer. These SNPs alter the function of the tumor-suppressor proteins produced by these genes, contributing to cancer development.

13.5.2 Gene Knockout and Knockdown Technologies

To better understand the function of specific genes, scientists often use gene knockout (inactivation of a gene) or knockdown (reduction of gene expression) techniques. In model organisms like mice, researchers can create knockout models where a gene is entirely removed from the genome, allowing them to observe the resulting phenotypic changes. Similarly, RNA interference (RNAi) techniques can be used to reduce the expression of specific genes in cells or organisms, providing insight into their roles. These techniques have been particularly valuable in studying genes with unknown functions. By analyzing the phenotypic consequences of knocking out a gene, scientists can infer its role in cellular processes or in the development of diseases.

13.6 MOVING TOWARD FUNCTIONAL GENOMICS

Functional genomics aims to provide a comprehensive understanding of how genes and their products work together within a living organism. The ultimate goal is to link every gene in the genome to its functional role in the body. While significant progress has been made, especially with the advent of high-throughput sequencing, much work remains to be done. As new technologies emerge, such as CRISPR-Cas9 for gene editing and single-cell RNA sequencing for detailed transcriptomic analysis, researchers are gaining increasingly precise insights into gene function. These advances are enabling a more complete understanding of how our genomes shape our biology, how genetic diseases arise, and how we can develop targeted therapies to treat them.

13.7 CONCLUSION

Understanding gene function is a complex process that goes beyond reading the genome sequence. It requires the integration of genomic, transcriptomic, proteomic, and phenotypic data to build a comprehensive understanding of how genes contribute to life. From transcriptional regulation and RNA processing to protein synthesis and genetic variation, each step in the journey from genome sequence to gene function adds a layer of complexity. However, with the continuous advancement of technologies and methodologies, we are progressively uncovering the mechanisms that link genetic information to biological outcomes, with vast implications for medicine, biotechnology, and beyond.

PART 4:

HUMAN GENOMICS

CHAPTER 14
FINDING HUMAN DISEASE GENES

14.1 INTRODUCTION: THE QUEST FOR DISEASE GENES

Understanding the genetic basis of human diseases has been a major focus of scientific research for decades. Identifying the genes responsible for various inherited disorders, as well as the genetic factors that influence common diseases like cancer, diabetes, and cardiovascular disease, is essential for improving diagnosis, treatment, and prevention. Advances in genomic technologies, such as high-throughput sequencing and genome-wide association studies (GWAS), have revolutionized our ability to pinpoint disease genes.

14.2 THE IMPORTANCE OF IDENTIFYING DISEASE GENES

Human diseases can be caused by mutations in single genes, known as monogenic diseases, or by the interaction of multiple genes and environmental factors, which contribute to polygenic diseases. Identifying the specific genes responsible for these conditions provides insights into their molecular mechanisms and allows researchers to develop more targeted therapies and diagnostic tools. Moreover, finding disease genes has profound implications for genetic counseling, prenatal testing, and personalized medicine. Understanding the genetic causes of disease can help healthcare providers offer better risk assessments and treatment plans tailored to an individual's genetic makeup.

14.3 APPROACHES TO FINDING DISEASE GENES

The process of identifying disease genes typically follows several key methodologies. These include linkage analysis, candidate gene studies, positional cloning, and more recent techniques such as GWAS and whole-genome sequencing. Each approach offers unique advantages and challenges, and their effectiveness depends on the nature of the disease being studied.

14.3.1 Linkage Analysis

Linkage analysis was one of the earliest methods used to identify disease genes, particularly for monogenic disorders. This technique involves studying families with inherited diseases and tracking how certain genetic markers are inherited alongside the disease trait. By observing which genetic markers are frequently passed on with the disease, scientists can narrow down the region of the genome where the disease gene is located. Linkage analysis works best for single-gene diseases that follow Mendelian inheritance patterns. For example, it was instrumental in identifying the gene responsible for Huntington's disease, a neurodegenerative disorder caused by a mutation in the HTT gene. However, the resolution of linkage analysis is limited, and it can be challenging to apply to complex diseases involving multiple genes or environmental factors.

14.3.2 Candidate Gene Studies

Candidate gene studies involve identifying genes that are suspected to be involved in a particular disease based on their known functions, biological pathways, or previous research. Researchers then examine the genetic variation within these candidate genes in individuals with and without the disease to identify specific mutations or variants that are associated with the condition.

For example, researchers studying cystic fibrosis, a genetic disorder that affects the lungs and digestive system, focused on genes related to ion transport. This led to the identification of the CFTR gene, mutations of which are responsible for the disease. While candidate gene studies have been successful in identifying genes involved in certain diseases, their scope is limited by the researcher's understanding of the disease's underlying biology.

14.3.3 Positional Cloning

Positional cloning is a more advanced approach used to identify disease genes, particularly for conditions where the genetic causes are not yet known. This method involves identifying a chromosomal region associated with the disease through linkage analysis or other mapping techniques, and then systematically narrowing down the region to locate the causative gene. In positional cloning, researchers may look for specific genetic markers in the region of interest and sequence the candidate genes in that region. This approach was instrumental in identifying the gene responsible for Duchenne muscular dystrophy (DMD), a genetic disorder that causes progressive muscle degeneration. However, positional cloning can be time-consuming and challenging, especially when the disease gene is located in a large genomic region.

14.4 HIGH-THROUGHPUT TECHNIQUES: REVOLUTIONIZING DISEASE GENE DISCOVERY

Recent advances in genomics have enabled researchers to identify disease genes more efficiently and accurately. Techniques such as genome-wide association studies (GWAS), whole-exome sequencing (WES), and whole-genome sequencing (WGS) have significantly expanded the scope of genetic research and have accelerated the discovery of genes involved in complex diseases.

14.4.1 Genome-Wide Association Studies (GWAS)

GWAS is a powerful technique used to identify genetic variants associated with complex diseases. Unlike earlier methods that focused on specific candidate genes, GWAS scans the entire genome to look for associations between common genetic variants (such as single nucleotide polymorphisms, or SNPs) and diseases.

By comparing the genomes of large numbers of individuals with and without a particular disease, GWAS can identify regions of the genome that are more likely to contain disease-associated genes. GWAS has been particularly successful in identifying risk factors for complex diseases like heart disease, diabetes, and cancer. For example, GWAS has revealed numerous genetic loci associated with Alzheimer's disease, although the specific genes and their functional roles are still being studied.

While GWAS has led to significant discoveries, it typically identifies common variants that have small effects on disease risk. As a result, these studies are often more effective at identifying disease risk factors rather than pinpointing causative mutations.

14.4.2 Whole-Genome and Whole-Exome Sequencing

Whole-genome sequencing (WGS) and whole-exome sequencing (WES) are powerful tools for identifying disease-causing mutations. WGS involves sequencing the entire genome, including both coding (exons) and non-coding regions, whereas WES focuses only on the exonic regions, which contain the protein-coding portions of genes. These approaches provide a comprehensive view of an individual's genetic makeup and allow researchers to identify rare or novel mutations that might be missed by other techniques. For example, WES has been used to identify rare genetic variants associated with diseases such as intellectual disabilities, rare cancers, and inherited syndromes. In cases where a genetic disorder is suspected but the causative gene is unknown, WGS and WES can be used to identify novel mutations, even in cases where the disease follows an autosomal recessive or complex inheritance pattern.

14.5 CHALLENGES IN IDENTIFYING DISEASE GENES

14.5.1 Genetic Heterogeneity

Genetic heterogeneity refers to the presence of different mutations in different genes that can cause the same or similar disease. This is common in complex diseases, where multiple genetic variants can contribute to the disease phenotype. As a result, it can be challenging to pinpoint a single causative gene or variant, particularly in diseases like schizophrenia or autism, which have many genetic and environmental risk factors.

14.5.2 Environmental Factors

Many common diseases, such as heart disease, diabetes, and cancer, are influenced by a combination of genetic factors and environmental exposures. This makes it difficult to isolate the specific genetic components that contribute to the disease. In such cases, gene-environment interactions must be carefully studied to understand the full range of factors that contribute to disease development.

14.5.3 Rare Variants and Ethical Issues

While common variants are relatively easy to identify through GWAS and other large-scale techniques, rare variants often require sequencing of large numbers of individuals to detect. Rare variants can have strong effects on disease risk, but they are difficult to study due to their low frequency in the population. Additionally, ethical concerns regarding the use of genetic data for disease risk prediction, privacy, and informed consent remain important issues in disease gene discovery.

14.6 THE FUTURE OF DISEASE GENE DISCOVERY

The future of human disease gene discovery lies in continued technological advancements, improved understanding of gene-environment interactions, and the integration of multi-omics data. The development of tools like CRISPR-Cas9 for gene editing and advances in single-cell RNA sequencing will further enhance our ability to study disease at the molecular level. Moreover, as personalized medicine continues to evolve, the identification of disease genes will play an increasingly important role in providing tailored treatments that target the specific genetic causes of disease. This will not only improve patient outcomes but also pave the way for preventative strategies that reduce the risk of genetic diseases.

14.7 CONCLUSION

Finding human disease genes is a complex and evolving process that requires a combination of genetic techniques, computational methods, and an understanding of disease biology. From the early methods of linkage analysis to the current use of high-throughput sequencing technologies, significant progress has been made in uncovering the genetic underpinnings of human diseases. Despite challenges such as genetic heterogeneity and gene-environment interactions, the future of disease gene discovery holds great promise, offering new opportunities for better diagnostics, treatments, and personalized medicine.

CHAPTER 15

UNDERSTANDING THE GENETIC BASIS OF CANCER

Cancer is a complex group of diseases characterized by uncontrolled cell growth and spread to other parts of the body. The underlying causes of cancer are multifactorial, involving genetic mutations, epigenetic changes, and environmental influences. The genetic basis of cancer is particularly important because it provides insights into how cancer develops, how it can be diagnosed, and what treatments may be most effective. Over the past few decades, advances in genomics and molecular biology have drastically improved our understanding of the genetic alterations that lead to cancer.

15.1 CANCER

Cancer arises from the accumulation of genetic alterations in normal cells that allow them to bypass the mechanisms that normally regulate growth and division. In a healthy cell, the genetic material, or DNA, is tightly regulated to maintain homeostasis. However, mutations in certain genes can disrupt this regulation, leading to uncontrolled cell division, invasion of surrounding tissues, and the potential for metastasis, or the spread of cancer to other parts of the body. These mutations can occur in two primary classes of genes: **oncogenes** and **tumor suppressor genes.** Oncogenes promote cell growth and division, while tumor suppressor genes normally inhibit cell growth and promote repair mechanisms. Mutations in these genes can tip the balance of cell growth, resulting in cancer.

15.2 ONCOGENES: THE GAS PEDALS OF CANCER

An oncogene is a gene that has the potential to cause cancer. In tumor cells, these genes are often mutated, or expressed at high levels. Most normal cells undergo a preprogrammed rapid cell death (apoptosis) if critical functions are altered and then malfunction. Activated oncogenes can cause those cells designated for apoptosis to survive and proliferate instead. Most oncogenes began as proto-oncogenes: normal genes involved in cell growth and proliferation or inhibition of apoptosis. If, through mutation, normal genes promoting cellular growth are up-regulated (gain-of-function mutation), they predispose the cell to cancer and are termed *oncogenes*. Usually, multiple oncogenes, along with mutated apoptotic or tumor suppressor genes, act in concert to cause cancer. Since the 1970s, dozens of oncogenes have been identified in human cancer. Many cancer drugs target the proteins encoded by oncogenes.

Oncogenes are a physically and functionally diverse set of genes, and as a result, their protein products have pleiotropic effects on a variety of intricate regulatory cascades within the cell.

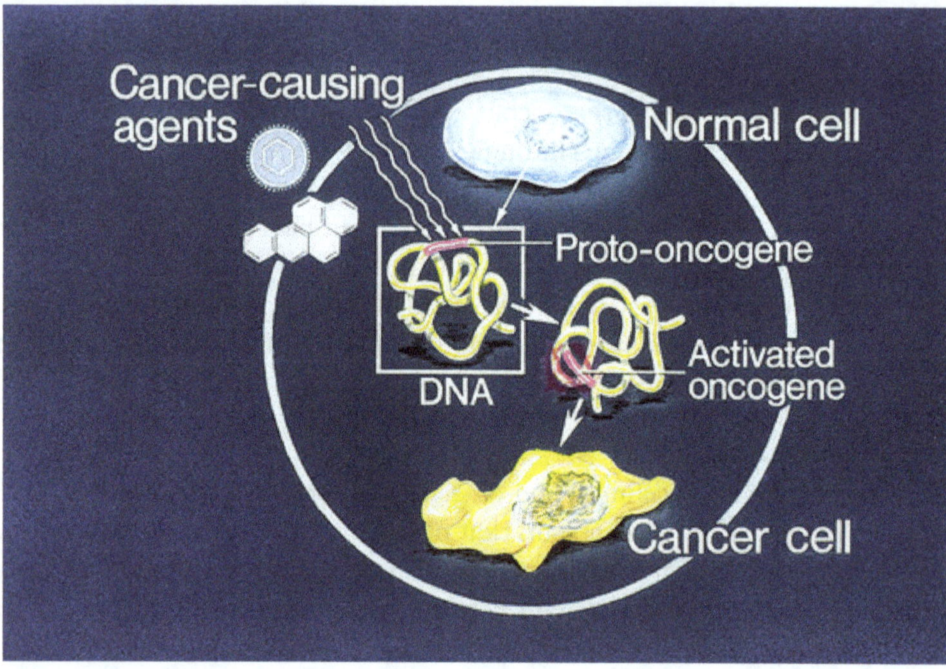

Fig: Illustration of how a Normal Cell is Converted to a Cancer Cell, when an Oncogene becomes Activated

Genes known as proto-oncogenes are those that normally encourage cell growth and division in order to generate new cells or sustain the viability of pre-existing cells. When overexpressed, proto-oncogenes can be inadvertently activated (turned on), which changes them to oncogenes.

There are numerous ways to activate (turn on) oncogenes in cells:

- **Gene Changes or Mutations:** A person's genetic "coding" may differ in a way that causes an oncogene to always be activated. These types of gene changes can develop spontaneously throughout the course of a person's life or they might be inherited from a parent when a transcription error occurs during cell division. Cells can frequently switch genes on or off via epigenetic mechanisms rather than actual genetic alterations. Alternately, different chemical compounds that can be linked to genetic material (DNA or RNA) may have an impact on which genes are active. An oncogene may sporadically become activated due to these epigenetic modifications. Visit Gene Alterations and Cancer to learn more about epigenetic alterations.

- **Chromosomal Rearrangement:** Every living creature has chromosomes, which are substantial strands of DNA that contain the genes for a cell. A chromosome's DNA sequence may alter each time a cell divides. This could cause a gene to be located near to a proto-oncogene that acts as an "on" switch, keeping it active even when it shouldn't. The cell can develop irregularly with the aid of this new oncogene.
- **Gene Duplication:** If one cell has more copies of a gene than another, that cell may produce too much of a certain protein.

The first human oncogene (HRAS), a crucial finding in the field of cancer research, was discovered more than 40 years ago, and since then, the number of novel pathogenic oncogenes has increased steadily. The discovery of specific small-molecule inhibitors that specifically target the different oncogenic proteins and a comprehensive mechanistic analysis of the ways in which oncogenes dysregulate physiological signaling to cause different cancer types and developmental syndromes are potential future advances in the field of cancer research. Investigating the quickly expanding field of oncogene molecular research, the goal of this special issue was to generate practical translational indicators that could be able to meet clinical needs.

Genes that are considered crucial for cancer can be divided into two categories based on whether the harmful mutations in them result in function loss or gain. Gain-of-function mutations of proto-oncogenes drive cells to proliferate when they shouldn't, while loss-of-function mutations of tumor suppressor genes free cells from inhibitions that typically serve to control their numbers. The ability of the mutant genes, known as oncogenes, to steer a specific line of test cells toward malignant proliferation can occasionally be used to identify these later mutations, which have a dominating effect.

Many of them were initially found to induce cancer in animals when they are introduced through viral vector infection, which carries genetic information from a prior host cell. Another method for identifying oncogenes is to look for genes that are activated by mutations in human cancer cells or by chromosomal translocations that may indicate the presence of a gene that is crucial for cancer.

Cancer patients are generally categorized according to clinical parameters in order to tailor their cancer therapy. For example, the separation of patients with acute leukemia into those with lymphocytic leukemia and those with myelocytic leukemia is important, because the optimal treatment for each form is different. Even in a particular disease, the identification of patients with good and poor prognostic potential is helpful, since more aggressive therapy may be needed to achieve a cure in the poor prognostic group. Oncogenes are prognostic markers in certain human cancers. N-myc amplification is an independent determinant in predicting a poor outcome in childhood neuroblastoma. Those children with amplification of N-myc, regardless of stage, will have shortened survival. Thus, therapeutic efforts are concentrated on intensifying treatment in this poor prognostic group.

15.3 TUMOR SUPPRESSOR GENES: THE BRAKES OF CANCER

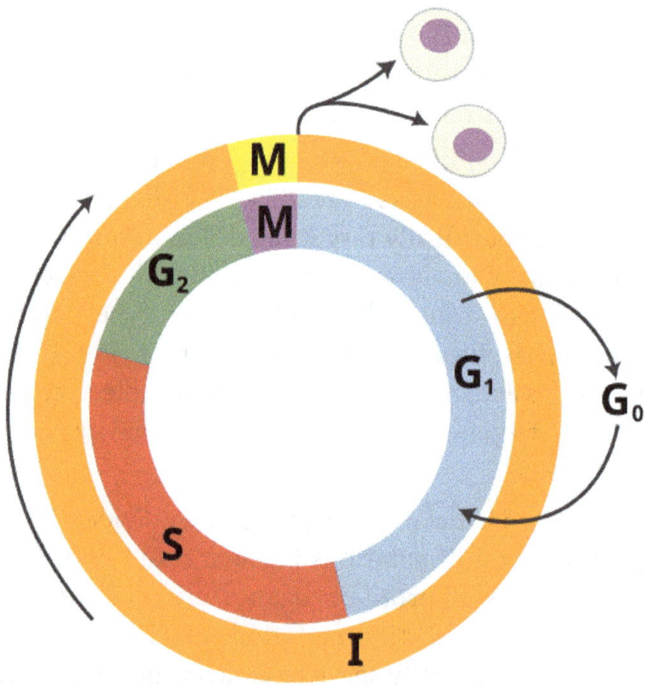

Fig: The Cell Cycle

A tumor suppressor gene (TSG), or anti-oncogene, is a gene that regulates a cell during cell division and replication. If the cell grows uncontrollably, it will result in cancer. When a tumor suppressor gene is mutated, it results in a loss or reduction in its function. In combination with other genetic mutations, this could allow the cell to grow abnormally. The loss of function for these genes may be even more significant in the development of human cancers, compared to the activation of oncogenes.

TSGs can be grouped into the following categories: caretaker genes, gatekeeper genes, and more recently landscaper genes. Caretaker genes ensure stability of the genome via DNA repair and subsequently when mutated allow mutations to accumulate. Meanwhile, gatekeeper genes directly regulate cell growth by either inhibiting cell cycle progression or inducing apoptosis. Lastly, landscaper genes regulate growth by contributing to the surrounding environment, and when mutated, can cause an environment that promotes unregulated proliferation. The classification schemes are evolving as medical advances are being made from fields including molecular biology, genetics, and epigenetics.

15.3.1 History

The discovery of oncogenes and their ability to deregulate cellular processes related to cell proliferation and development appeared first in the literature as opposed to the idea of tumor suppressor genes. However, the idea of genetic mutation leading to increased tumor growth gave way to another possible genetic idea of genes playing a role in decreasing cellular growth and development of cells. This idea was not solidified until experiments by Henry Harris were conducted with somatic cell hybridization in 1969.

Within Harris's experiments, tumor cells were fused with normal somatic cells to make hybrid cells. Each cell had chromosomes from both parents and upon growth, a majority of these hybrid cells did not have the capability of developing tumors within animals. The suppression of tumorigenicity in these hybrid cells prompted researchers to hypothesize that genes within the normal somatic cell had inhibitory actions to stop tumor growth. This initial hypothesis eventually lead to the discovery of the first classic tumor suppressor gene by Alfred Knudson, known as the Rb gene, which codes for the retinoblastoma tumor suppressor protein.

Alfred Knudson, a pediatrician and cancer geneticist, proposed that in order to develop retinoblastoma, two allelic mutations are required to lose functional copies of both the Rb genes to lead to tumorigenicity. Knudson observed that retinoblastoma often developed early in life for younger patients in both eyes, while in some rarer cases retinoblastoma would develop later in life and only be unilateral. This unique development pattern allowed Knudson and several other scientific groups in 1971 to correctly hypothesize that the early development of retinoblastoma was caused by inheritance of one loss of function mutation to an RB germ-line gene followed by a later de novo mutation on its functional Rb gene allele. The more sporadic occurrence of unilateral development of retinoblastoma was hypothesized to develop much later in life due to two de novo mutations that were needed to fully lose tumor suppressor properties. This finding formed the basis of the two-hit hypothesis. In order to verify that the loss of function of tumor suppressor genes causes increased tumorigenicity, interstitial deletion experiments on chromosome 13q14 were conducted to observe the effect of deleting the loci for the Rb gene. This deletion caused increased tumor growth in retinoblastoma, suggesting that loss or inactivation of a tumor suppressor gene can increase tumorigenicity.

15.3.2 Two-Hit Hypothesis

Unlike oncogenes, tumor suppressor genes generally follow the two-hit hypothesis, which states both alleles that code for a particular protein must be affected before an effect is manifested. If only one allele for the gene is damaged, the other can still produce enough of the correct protein to retain the appropriate function. In other words, mutant tumor suppressor alleles are usually recessive, whereas mutant oncogene alleles are typically dominant.

Proposed by A.G. Knudson for cases of retinoblastoma. He observed that 40% of U.S cases were caused by a mutation in the germ-line. However, affected parents could have children without the disease, but the unaffected children became parents of children with retinoblastoma. This indicates that one could inherit a mutated germ-line but not display the disease. Knudson observed that the age of onset of retinoblastoma followed 2nd order kinetics, implying that two independent genetic events were necessary.

He recognized that this was consistent with a recessive mutation involving a single gene, but requiring bi-allelic mutation. Hereditary cases involve an inherited mutation and a single mutation in the normal allele. Non-hereditary retinoblastoma involves two mutations, one on each allele. Knudson also noted that hereditary cases often developed bilateral tumors and would develop them earlier in life, compared to non-hereditary cases where individuals were only affected by a single tumor.

There are exceptions to the two-hit rule for tumor suppressors, such as certain mutations in the p53 gene product. p53 mutations can function as a dominant negative, meaning that a mutated p53 protein can prevent the function of the natural protein produced from the non-mutated allele. Other tumor-suppressor genes that do not follow the two-hit rule are those that exhibit haploinsufficiency, including PTCH in medulloblastoma and NF1 in neurofibroma. Another example is p27, a cell-cycle inhibitor, that when one allele is mutated causes increased carcinogen susceptibility.

15.3.3 Functions

The proteins encoded by most tumor suppressor genes inhibit cell proliferation or survival. Inactivation of tumor suppressor genes therefore leads to tumor development by eliminating negative regulatory proteins. In most cases, tumor suppressor proteins inhibit the same cell regulatory pathways that are stimulated by the products of oncogenes. While tumor suppressor genes have the same main function, they have various mechanisms of action, that their transcribed products perform, which include the following:

- **Intracellular Proteins:** That control gene expression of a specific stage of the cell cycle. If these genes are not expressed, the cell cycle does not continue, effectively inhibiting cell division.

- **Receptors or Signal Transducers:** Secreted hormones or developmental signals that inhibit cell proliferation.
- **Checkpoint-Control Proteins:** That trigger cell cycle arrest in response to DNA damage or chromosomal defects.
- **Proteins that Induce Apoptosis:** If damage cannot be repaired, the cell initiates programmed cell death to remove the threat it poses to the organism as a whole..
- **Cell Adhesion:** Some proteins involved in cell adhesion prevent tumor cells from dispersing, block loss of contact inhibition, and inhibit metastasis. These proteins are known as metastasis suppressors.
- **Proteins Involved in Repairing Mistakes in DNA:** Caretaker genes encode proteins that function in repairing mutations in the genome, preventing cells from replicating with mutations. Furthermore, increased mutation rate from decreased DNA repair leads to increased inactivation of other tumor suppressors and activation of oncogenes.
- **Certain Genes can also Act as Tumor Suppressors and Oncogenes:** Dubbed Proto-oncogenes with Tumor suppressor function, these genes act as "double agents" that both positively and negatively regulate transcription.

15.3.4 Epigenetic Influences

Expression of genes, including tumor suppressors, can be altered through biochemical alterations known as DNA methylation. Methylation is an example of epigenetic modifications, which commonly regulate expression in mammalian genes. The addition of a methyl group to either histone tails or directly on DNA causes the nucleosome to pack tightly together restricting the transcription of any genes in this region. This process not only has the capabilities to inhibit gene expression, it can also increase the chance of mutations. Stephen Baylin observed that if promoter regions experience a phenomenon known as hypermethylation, it could result in later transcriptional errors, tumor suppressor gene silencing, protein misfolding, and eventually cancer growth. Baylin et al. found methylation inhibitors known as azacitidine and decitabine. These compounds can actually help prevent cancer growth by inducing re-expression of previously silenced genes, arresting the cell cycle of the tumor cell and forcing it into apoptosis.

There are further clinical trials under current investigation regarding treatments for hypermethylation as well as alternate tumor suppression therapies that include preventionoftissuehyperplasia,tumordevelopment,ormetastaticspreadoftumors. The team working with Wajed have investigated neoplastic tissue methylation in order to one day identify early treatment options for gene modification that can silence the tumor suppressor gene. In addition to DNA methylation, other epigenetic modifications like histone deacetylation or chromatin-binding proteins can prevent DNA polymerase from effectively transcribing desired sequences, such as ones containing tumor suppressor genes.

15.3.5 Clinical Significance

Gene therapy is used to reinstate the function of a mutated or deleted gene type. When tumor suppressor genes are altered in a way that results in less or no expression, several severe problems can arise for the host. This is why tumor suppressor genes have commonly been studied and used for gene therapy. The two main approaches used currently to introduce genetic material into cells are viral and non-viral delivery methods.

Viral Methods

The viral method of transferring genetic material harnesses the power of viruses. By using viruses that are durable to genetic material alterations, viral methods of gene therapy for tumor suppressor genes have shown to be successful. In this method, vectors from viruses are used. The two most commonly used vectors are adenoviral vectors and adeno-associated vectors. In vitro genetic manipulation of these types of vectors is easy and in vivo application is relatively safe compared to other vectors. Before the vectors are inserted into the tumors of the host, they are prepared by having the parts of their genome that control replication either mutated or deleted. This makes them safer for insertion. Then, the desired genetic material is inserted and ligated to the vector. In the case with tumor suppressor genes, genetic material which encodes p53 has been used successfully, which after application, has shown reduction in tumor growth or proliferation.

Non-Viral Methods

The non-viral method of transferring genetic material is used less often than the viral method. However, the non-viral method is a more cost-effective, safer, available method of gene delivery not to mention that non-viral methods have shown to induce fewer host immune responses and possess no restrictions on size or length of the transferable genetic material. Non-viral gene therapy uses either chemical or physical methods to introduce genetic material to the desired cells. The chemical methods are used primarily for tumor suppressor gene introduction and are divided into two categories which are naked plasmid or liposome-coated plasmids. The naked plasmid strategy has garnered interest because of its easy to use methods. Direct injection into the muscles allows for the plasmid to be taken up into the cell of possible tumors where the genetic material of the plasmid can be incorporated into the genetic material of the tumor cells and revert any previous damage done to tumor suppressor genes.

The liposome-coated plasmid method has recently also been of interest since they produce relatively low host immune response and are efficient with cellular targeting. The positively charged capsule in which the genetic material is packaged helps with electrostatic attraction to the negatively charged membranes of the cells as well as the negatively charged DNA of the tumor cells. In this way, non-viral methods of gene therapy are highly effective in restoring tumor suppressor gene function to tumor cells that have either partially or entirely lost this function.

Limitations

The most important limitation these methods have is the efficacy at which the adenoviral and adeno-associated vectors, naked plasmids, or liposome-coated plasmids are taken in by the host's tumor cells.

If proper uptake by the host's tumor cells is not achieved, re-insertion introduces problems such as the host's immune system recognizing these vectors or plasmids and destroying them which impairs the overall effectiveness of the gene therapy treatment further.

15.3.6 Examples

Gene	Original Function	Two-Hit	Associated Carcinomas
Rb	DNA Replication, cell division and death	Yes	Retinoblastoma
p53	Apoptosis	No	Half of all known malignancies
VHL	Cell division, death, and differentiation	Yes	Kidney Cancer
APC	DNA damage, cell division, migration, adhesion, death	Yes	Colorectal Cancer
BRCA2	Cell division and death, and repair of double-stranded DNA breaks	Yes	Breast/Ovarian Cancer
NF1	Cell differentiation, division, development, RAS signal transduction	No	Nerve tumors, Neuroblastoma
PTCH	Hedgehog signaling	No	Medulloblastoma, Basal Cell Carcinoma

- **Retinoblastoma Protein (pRb):** pRb was the first tumor-suppressor protein discovered in human retinoblastoma; however, recent evidence has also implicated pRb as a tumor-survival factor. *RB1* gene is a gatekeeper gene that blocks cell proliferation, regulates cell division and cell death. Specifically pRb prevents the cell cycle progression from G1 phase into the S phase by binding to E2F and repressing the necessary gene transcription. This prevents the cell from replicating its DNA if there is damage.

- **p53:** *TP53*, a caretaker gene, encodes the protein p53, which is nicknamed "the guardian of the genome". p53 has many different functions in the cell including DNA repair, inducing apoptosis, transcription, and regulating the cell cycle. Mutated p53 is involved in many human cancers, of the 6.5 million cancer diagnoses each year about 37% are connected to p53 mutations. This makes it a popular target for new cancer therapies. Homozygous loss of p53 is found in 65% of colon cancers, 30–50% of breast cancers, and 50% of lung cancers. Mutated p53 is also involved in the pathophysiology of leukemias, lymphomas, sarcomas, and neurogenic tumors. Abnormalities of the p53 gene can be inherited in Li-Fraumeni syndrome (LFS), which increases the risk of developing various types of cancers.
- **BCL2:** BCL2 is a family of proteins that are involved in either inducing or inhibiting apoptosis. The main function is involved in maintaining the composition of the mitochondria membrane, and preventing cytochrome c release into the cytosol. When cytochrome c is released from the mitochondria it starts a signaling cascade to begin apoptosis.
- **SWI/SNF:** SWI/SNF is a chromatin remodeling complex, which is lost in about 20% of tumors. The complex consists of 10-15 subunits encoded by 20 different genes. Mutations in the individual complexes can lead to misfolding, which compromises the ability of the complex to work together as a whole. SWI/SNF has the ability move nucleosomes, which condenses DNA, allowing for transcription or block transcription from occurring for certain genes. Mutating this ability could cause genes to be turned on or off at the wrong times.

15.4 THE ROLE OF DNA REPAIR IN CANCER

The ability of cells to maintain and repair their DNA is crucial for preventing cancer. DNA damage can arise from environmental factors like UV radiation, smoking, and certain chemicals, or from normal cellular processes like replication errors. DNA repair mechanisms, such as mismatch repair, nucleotide excision repair, and double-strand break repair, are essential for maintaining genome integrity. Mutations in genes responsible for DNA repair can predispose individuals to cancer. For example, **Lynch syndrome**, a hereditary cancer syndrome, is caused by mutations in mismatch repair genes, leading to an increased risk of colorectal, endometrial, and other cancers. Similarly, defects in the DNA repair pathway can contribute to the development of cancer in individuals with **Xeroderma pigmentosum**, a rare condition that increases susceptibility to skin cancer due to a defect in the nucleotide excision repair pathway.

15.5 EPIGENETICS AND CANCER

In addition to genetic mutations, **epigenetic changes**—which refer to modifications that affect gene expression without altering the underlying DNA sequence—also play a key role in cancer. Epigenetic alterations can regulate whether certain genes are turned on or off, and when these changes affect tumor suppressor genes or oncogenes, they can contribute to cancer.

One of the most common epigenetic changes in cancer is **DNA methylation**, where methyl groups are added to the DNA molecule, typically silencing gene expression. Tumor suppressor genes like p16INK4a and MLH1 can be silenced by abnormal DNA methylation in cancer cells, contributing to the loss of control over cell division. Another important epigenetic mechanism is **histone modification**, where chemical changes to histone proteins affect the packaging of DNA in the nucleus and influence gene expression. Abnormal histone modifications can disrupt the regulation of genes involved in cell growth and differentiation, contributing to tumorigenesis.

15.6 THE ROLE OF INHERITED MUTATIONS IN CANCER

While most cancers are caused by sporadic genetic mutations that occur during a person's lifetime, some individuals inherit mutations that increase their risk of developing certain types of cancer.

These inherited mutations typically involve oncogenes or tumor suppressor genes, and they can be passed down from one generation to the next. Inherited mutations in the **BRCA1** and **BRCA2** genes significantly increase the risk of breast and ovarian cancer. Similarly, mutations in the **RET** gene are associated with medullary thyroid cancer, and mutations in **MLH1** and **MSH2** can lead to Lynch syndrome. Family history and genetic testing are crucial tools in identifying individuals with inherited cancer risk, enabling early detection and preventive strategies.

15.7 THE ROLE OF THE MICROENVIRONMENT IN CANCER DEVELOPMENT

The development of cancer is not solely determined by the genetic alterations within cancer cells. The **tumor microenvironment**—which includes surrounding tissues, blood vessels, immune cells, and extracellular matrix—also plays a critical role in cancer progression. The interactions between tumor cells and their microenvironment can influence tumor growth, metastasis, and response to treatment. For example, tumors can recruit blood vessels through a process called

angiogenesis, allowing them to receive more oxygen and nutrients. Tumor cells can also manipulate the immune system to evade immune surveillance, making it harder for the body to recognize and destroy cancer cells.

15.8 ADVANCES IN CANCER GENETICS AND TREATMENT

Understanding the genetic basis of cancer has led to significant advances in cancer treatment, particularly through the development of **targeted therapies** and **personalized medicine.** Targeted therapies are drugs that specifically target the genetic mutations or pathways driving cancer, minimizing damage to normal cells and reducing side effects.

For example, **Herceptin** (trastuzumab) is a monoclonal antibody used to treat breast cancers that overexpress the HER2 protein. Similarly, **immunotherapies** like checkpoint inhibitors aim to activate the immune system to target cancer cells more effectively.

Personalized medicine, which tailors treatment plans based on a patient's genetic profile, is becoming an increasingly important aspect of cancer treatment. By identifying specific mutations in a patient's cancer cells, doctors can choose therapies that are more likely to be effective, improving treatment outcomes.

15.9 CONCLUSION

The genetic basis of cancer is an evolving field, with continuous advancements in technology and research leading to more precise and effective treatments. Understanding the genetic alterations that drive cancer, from oncogenes to tumor suppressor genes, as well as the role of DNA repair and epigenetics, is crucial for developing targeted therapies. With the advent of personalized medicine and genetic testing, the future of cancer treatment holds the promise of more tailored and effective interventions, improving outcomes for patients around the world.

CHAPTER 16

DNA FINGERPRINTING AND FORENSICS

DNA fingerprinting, also known as DNA profiling, is a technique used to identify individuals based on their unique DNA characteristics. Unlike traditional forms of identification, such as fingerprints or facial recognition, DNA fingerprinting offers a highly precise method for distinguishing between individuals. This method relies on the fact that the DNA of each person, with the exception of identical twins, is unique. In forensic science, DNA fingerprinting has revolutionized the field by providing an invaluable tool for solving crimes, exonerating the innocent, and identifying victims in situations where visual identification is impossible.

16.1 THE SCIENCE BEHIND DNA FINGERPRINTING

DNA fingerprinting is based on the principle that certain regions of the human genome contain highly variable sequences, known as **short tandem repeats (STRs)** or **variable number tandem repeats (VNTRs)**. These regions are highly polymorphic, meaning they vary greatly between individuals, and are located in non-coding areas of the genome. While the majority of the human genome consists of coding regions that determine traits, these non-coding regions do not have a direct function in the organism's development, making them ideal markers for identification purposes.

When a forensic scientist performs DNA fingerprinting, they extract DNA from a biological sample such as blood, hair, or saliva. The DNA is then processed and analyzed using specific molecular biology techniques like **polymerase chain reaction (PCR)** and **gel electrophoresis**. PCR amplifies the regions of interest, and gel electrophoresis is used to separate and visualize the DNA fragments based on their size. The resulting pattern of bands is compared with that from other samples, creating a DNA "fingerprint" unique to an individual.

16.2 DNA FINGERPRINTING AND ITS APPLICATION IN FORENSICS

DNA fingerprinting has become a cornerstone in forensic science, used in a variety of applications ranging from criminal investigations to disaster victim identification. It provides highly reliable evidence in the form of DNA samples that can be matched to suspects, victims, or crime scenes. One of the key benefits of

DNA fingerprinting in forensics is its ability to generate conclusive evidence even from small or degraded samples.

16.2.1 Crime Scene Investigation

In criminal investigations, DNA evidence is crucial for linking a suspect to a crime scene or ruling out potential suspects. For example, DNA found at a crime scene, such as blood, hair, or skin cells, can be compared to the DNA profile of a suspect. If the profiles match, it provides strong evidence that the individual was present at the scene of the crime. Moreover, DNA evidence can also be used to establish a suspect's exclusion from a crime scene. In many cases, forensic experts use DNA to rule out individuals who may have been wrongly accused of a crime, providing a level of justice that traditional evidence may not.

16.2.2 Identification of Victims

Another significant application of DNA fingerprinting is in the identification of victims, especially in mass disasters such as plane crashes, natural disasters, or terrorist attacks. In such cases, the body may be so damaged that traditional methods of identification, like dental records or visual recognition, are not feasible. DNA from remains can be compared to DNA samples from family members to confirm the identity of the deceased, offering closure to families in situations of tragedy.

16.2.3 Paternity Testing

DNA fingerprinting has also become widely used in the context of family law, particularly in paternity testing. By comparing the DNA profiles of a mother, child, and potential father, it is possible to determine biological relationships with a high degree of accuracy. This type of DNA analysis has significant implications for child support, inheritance claims, and custody disputes.

16.2.4 Cold Case Investigations

One of the most powerful uses of DNA fingerprinting is in the resolution of cold cases, where crimes have remained unsolved for years or even decades. As DNA evidence can be preserved for long periods, it allows investigators to reexamine old evidence from cases that may have previously lacked sufficient leads. Advances in DNA technology have also made it possible to extract usable DNA from degraded or compromised samples. Over the years, numerous cold cases have been solved by revisiting old evidence with modern DNA techniques.

16.3 ETHICAL AND LEGAL CONSIDERATIONS IN DNA FINGERPRINTING

16.3.1 Privacy Concerns

One of the primary ethical concerns surrounding DNA fingerprinting is privacy. As DNA profiling can reveal sensitive personal information, such as genetic predispositions to certain diseases, there is a risk of this information being misused. DNA databases, which are increasingly used by law enforcement, may inadvertently lead to genetic profiling or discrimination based on an individual's genetic traits, leading to concerns about civil liberties and privacy violations.

16.3.2 The Reliability of DNA Evidence

Although DNA fingerprinting is highly reliable, errors can still occur, especially when samples are contaminated or mishandled. Forensic laboratories are required to follow strict protocols to avoid contamination, but even small mistakes can have serious consequences. The accuracy of DNA evidence in court can also be influenced by the quality of the samples and the laboratory's methods. Inaccurate or incomplete analyses can lead to wrongful convictions or the dismissal of evidence.

16.3.3 Legal Use of DNA Evidence

The legal system must ensure that DNA evidence is used fairly and appropriately. In some jurisdictions, individuals may be required to provide a DNA sample, even if they are not involved in a criminal investigation. While this helps law enforcement solve crimes, it raises concerns about individuals' rights to privacy and the potential for overreach. There must be clear and consistent laws governing how DNA evidence is collected, stored, and used in criminal investigations to prevent misuse or abuse.

16.4 ADVANCEMENTS IN DNA FINGERPRINTING TECHNOLOGY

The field of DNA fingerprinting has evolved significantly since its introduction in the 1980s. Early DNA profiling relied on the analysis of VNTRs, but modern techniques now focus on analyzing **short tandem repeats (STRs)**, which are shorter and more stable, making them easier to work with, especially in degraded samples. STR analysis involves examining specific regions of the DNA where a repeating sequence of two to six base pairs occurs. The number of repeats varies between individuals, providing a unique genetic fingerprint.

Advancements in **next-generation sequencing (NGS)** technologies have also expanded the capabilities of DNA fingerprinting. NGS allows for the sequencing of entire genomes, which can provide even more detailed genetic information. While NGS is not yet widely used in forensic DNA analysis due to cost and time constraints, its potential for improving forensic investigations is immense.

Another important development in the field is the use of **mitochondrial DNA (mtDNA) analysis**, particularly in cases where nuclear DNA is not available or is too degraded. Mitochondrial DNA is inherited maternally and can be found in higher quantities in cells than nuclear DNA. This makes mtDNA analysis particularly useful for identifying remains in cases where conventional DNA profiling is not possible.

16.5 CONCLUSION: THE ROLE OF DNA FINGERPRINTING IN MODERN FORENSICS

DNA fingerprinting has transformed forensic science, providing a powerful tool for criminal investigations, victim identification, and resolving family disputes. Its ability to generate precise and unique genetic profiles has enabled investigators to solve cases that would have otherwise remained unsolved, while also helping to exonerate individuals who were wrongly convicted. Despite the remarkable advancements in DNA fingerprinting technology, the ethical and legal considerations surrounding its use must be carefully managed to protect individuals' privacy and ensure the fairness of the judicial process.

As technology continues to evolve, the future of DNA fingerprinting in forensic science will likely see even more precise and efficient methods for analysis, with applications extending beyond traditional criminal cases to areas such as personalized medicine and ancestry research. However, ensuring that these advances are implemented ethically and responsibly remains a key challenge for both scientists and legal professionals alike.

BIBLIOGRAPHY

Advanced Genetic Analysis: Using Bioinformatics and Molecular Genetics: Authors: W.J. Dixon, Publisher: Oxford University Press, Year: 2008

Bacterial Pathogenesis and Immune Responses: A Molecular Approach: Authors: Margaret K. M. Ireton, Publisher: Springer, Year: 2014

Basic Biotechnology: Authors: Colin Ratledge, Bjorn Kristiansen, Publisher: Cambridge University Press, Year: 2006

Bioinformatics: Sequence and Genome Analysis: Authors: David W. Mount, Publisher: Cold Spring Harbor Laboratory Press, Year: 2004

Biotechnology: An Introduction: Authors: William J. Thieman, Michael A. Palladino, Publisher: Pearson, Year: 2014

Cloning and DNA Analysis: Authors: J. L. Kinsella, Publisher: Academic Press, Year: 2009

DNA Science: A First Course in Recombinant DNA Technology: Authors: David P. Clark, Publisher: Cold Spring Harbor Laboratory Press, Year: 2012

DNA Technology in Forensic Science: Authors: J. K. P. Fenton, Publisher: Springer, Year: 2015

Gene Cloning and DNA Analysis: An Introduction: Authors: T. A. Brown, Publisher: Wiley-Blackwell, Year: 2016

Gene Cloning and Gene Expression: The Basics: Authors: K. L. Simmons, Publisher: Springer, Year: 2013

Gene Therapy: Prospective Technology: Authors: K. S. G. Kartha, Publisher: Cambridge University Press, Year: 2001

Gene Therapy: Therapeutic Mechanisms and Biomedical Applications: Authors: M.A. Korman, Publisher: Springer, Year: 2006

Genetic Engineering in Medicine: Authors: J. P. Bourne, Publisher: Elsevier, Year: 2007

Genetic Engineering: Principles and Applications: Authors: C. J. T. Caves, Publisher: Wiley, Year: 2011

Genetic Engineering: Principles and Methods: Authors: Jane M. D. Jenson, Sydney B. Barron, Publisher: Springer, Year: 2007

Genomes: Authors: T.A. Brown, Publisher: Garland Science, Year: 2006

Genomics and Bioinformatics: An Introduction to Programming Tools for Life Scientists: Authors: K. S. Rajasekaran, Publisher: Wiley-Interscience, Year: 2012

Genomics and Bioinformatics: Authors: Jeffry W. Wood, Publisher: Wiley-Blackwell, Year: 2010

Genomics: A Very Short Introduction: Authors: Jonathan A. Slack, Publisher: Oxford University Press, Year: 2014

Introduction to Biotechnology: Authors: William D. Schroeder, Publisher: Prentice Hall, Year: 2010

Introduction to Genetic Analysis: Authors: Anthony J. F. Griffiths, Susan R. Wessler, Sean B. Carroll, John Doebley, Publisher: W. H. Freeman, Year: 2011

Introduction to Genetic Engineering: Authors: Desmond S. T. G. Harrington, Publisher: Academic Press, Year: 2007

Introduction to Molecular Genetics and Biotechnology: Authors: F. John O'Neill, Publisher: Springer, Year: 2006

Laboratory Manual in Molecular Biology and Genetics: Authors: Robert S. White, Publisher: Wiley-Liss, Year: 2006

Methods in Molecular Biology: Recombinant DNA Technology: Authors: F. M. Ausubel, Publisher: Springer, Year: 2007

Microbial Biotechnology: Fundamentals of Applied Microbiology: Authors: Alexander N. S., Publisher: Springer, Year: 2012

Molecular Biology and Biotechnology: Authors: R. R. M. Purvis, Publisher: Oxford, Year: 2006

Molecular Biology of the Gene, Volume 4: Authors: Watson et al., Publisher: Pearson Education, Year: 2013

Molecular Biology of the Gene: A Self-Teaching Guide: Authors: James D. Watson, Publisher: Wiley, Year: 2005

Molecular Biology of the Gene: Authors: James D. Watson, Tania A. Baker, Stephen P. Bell, Publisher: Pearson, Year: 2014

Molecular Biotechnology: An Introduction: Authors: Glick B. R., Publisher: Springer, Year: 2010

Molecular Biotechnology: Principles and Applications of Recombinant DNA: Authors: Bernard R. Glick, Jack J. Pasternak, Publisher: ASM Press, Year: 2009

Molecular Cloning: A Laboratory Manual, Volume 2: Authors: Michael R. Green, Joseph Sambrook, Publisher: Cold Spring Harbor Laboratory Press, Year: 2012

Molecular Genetics of Bacteria: Authors: Larry Snyder, Wendy Champness, Publisher: ASM Press, Year: 2013

Molecular Genetics: Authors: K. P. Harris, Publisher: Pearson, Year: 2015

Practical Genetic Engineering: Authors: B. K. A. Dixit, Publisher: Pearson Education, Year: 2013

Principles of Bioinformatics: Authors: P. L. Leung, Publisher: Pearson, Year: 2014

Principles of Gene Manipulation and Genomics: Authors: Sandy B. Primrose, Richard Twyman, Publisher: Wiley-Blackwell, Year: 2006

Recombinant DNA and Biotechnology: Authors: S. C. Harland, Publisher: Prentice Hall, Year: 2009

Recombinant DNA Technology: A Laboratory Manual: Authors: S. S. P. Hegde, L. S. R. K. Shankar, Publisher: Springer, Year: 2014

Recombinant DNA: A Laboratory Manual: Authors: Jeffrey A. Lee, C. M. Herrick, Publisher: McGraw-Hill, Year: 2006

Recombinant DNA: From Molecular Cloning to Genomic Medicine: Authors: Mark A. S. M. Demetriou, Publisher: Springer, Year: 2007

Recombinant DNA: Genes and Genomes - A Short Course: Authors: James D. Watson, Michael Gilman, Stephen P. Hughes, Richard H. Roberts, Publisher: W. H. Freeman, Year: 2007

Techniques in Genetic Engineering: Authors: C. M. Berns, Publisher: Wiley-Blackwell, Year: 2013

The Molecular Biology of the Cell: Authors: Alberts, Bruce, Alexander Johnson, Julian Lewis, Publisher: Garland Science, Year: 2002

INDEX

www.ingramcontent.com/pod-product-compliance
Lightning Source LLC
LaVergne TN
LVHW062224200926
848484LV00006B/1466
* 9 7 8 1 7 8 7 1 5 4 5 0 6 *